Springer Biographies

The books published in the Springer Biographies tell of the life and work of scholars, innovators, and pioneers in all fields of learning and throughout the ages. Prominent scientists and philosophers will feature, but so too will lesser known personalities whose significant contributions deserve greater recognition and whose remarkable life stories will stir and motivate readers. Authored by historians and other academic writers, the volumes describe and analyse the main achievements of their subjects in manner accessible to nonspecialists, interweaving these with salient aspects of the protagonists' personal lives. Autobiographies and memoirs also fall into the scope of the series.

Sanja Damjanovic · Volker Metag ·
Jurgen Schukraft · Hans Joachim Specht
Editors

Hans Joachim Specht

Scientist and Visionary

With a Foreword by Carlo Rubbia

Editors
Sanja Damjanovic
Physics
CERN
Geneva, Switzerland

Volker Metag
II. Physikalisches Institut
Justus-Liebig-Universität Gießen
Giessen, Germany

Jurgen Schukraft
Physics
CERN
Geneva, Switzerland

Hans Joachim Specht
Physikalisches Institut
Universität Heidelberg
Heidelberg, Germany

ISSN 2365-0613   ISSN 2365-0621 (electronic)
Springer Biographies
ISBN 978-3-031-92352-4   ISBN 978-3-031-92353-1 (eBook)
https://doi.org/10.1007/978-3-031-92353-1

© The Editor(s) (if applicable) and The Author(s) 2025. This book is an open access publication.

Open Access This book is licensed under the terms of the Creative Commons Attribution 4.0 International License (http://creativecommons.org/licenses/by/4.0/), which permits use, sharing, adaptation, distribution and reproduction in any medium or format, as long as you give appropriate credit to the original author(s) and the source, provide a link to the Creative Commons license and indicate if changes were made.
The images or other third party material in this book are included in the book's Creative Commons license, unless indicated otherwise in a credit line to the material. If material is not included in the book's Creative Commons license and your intended use is not permitted by statutory regulation or exceeds the permitted use, you will need to obtain permission directly from the copyright holder.
The use of general descriptive names, registered names, trademarks, service marks, etc. in this publication does not imply, even in the absence of a specific statement, that such names are exempt from the relevant protective laws and regulations and therefore free for general use.
The publisher, the authors and the editors are safe to assume that the advice and information in this book are believed to be true and accurate at the date of publication. Neither the publisher nor the authors or the editors give a warranty, expressed or implied, with respect to the material contained herein or for any errors or omissions that may have been made. The publisher remains neutral with regard to jurisdictional claims in published maps and institutional affiliations.

This Springer imprint is published by the registered company Springer Nature Switzerland AG
The registered company address is: Gewerbestrasse 11, 6330 Cham, Switzerland

If disposing of this product, please recycle the paper.

1965

2017

Hans Joachim Specht
Sixty Years of Physics—
The Fascination of Diversity

A Portrait of Diverse Excellence

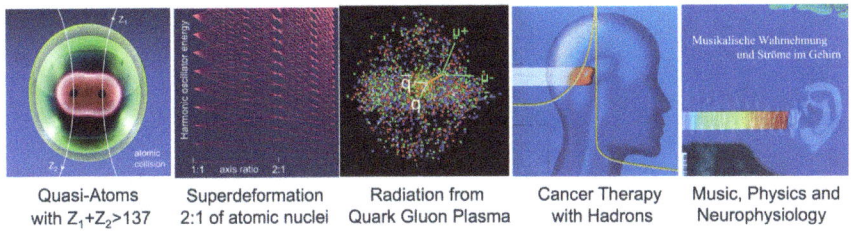

| Quasi-Atoms with $Z_1+Z_2>137$ | Superdeformation 2:1 of atomic nuclei | Radiation from Quark Gluon Plasma | Cancer Therapy with Hadrons | Music, Physics and Neurophysiology |

Hans Joachim Specht's illustrious scientific journey, where vision and diversity converge, has left a lasting impact on both science and society.

From discoveries in **atomic physics**, **nuclear fission**, and **ultra-relativistic heavy-ion physics** to pioneering **hadron cancer therapy**, and even exploring the **intersections between physics, music, and neuroscience**, Hans Joachim Specht's contribution to science and society stands as a testament to the boundless possibilities of scientific curiosity.

Foreword

In 1983, we discovered the W and Z bosons at the CERN Super Proton Synchrotron (SPS), and a year later, while the atmosphere was still one of celebration, the experimental ultra-relativistic heavy-ion program was launched at the CERN SPS, a pioneering effort in which Hans Specht, alongside Bill Willis, played a central role. Coming from the world of low-energy nuclear physics, Hans took on the challenge of becoming the spokesperson for CERN's first-generation heavy-ion experiment. Later, Hans set in motion and drove forward the second-generation experiment, CERES/NA45, an initiative that many doubted would ever succeed. I did, however, believe in it, and I personally supported the experiment, aimed at providing evidence for the existence of a new state of matter, the quark–gluon plasma, and I ensured its realization during my tenure as CERN's Director-General. The CERES results stand as one of the most cited achievements in SPS heavy-ion physics.

Hans's contributions to fundamental physics went far beyond the laboratory. He was a transformative leader at the GSI Helmholtz Center for Heavy Ion Research, elevating its international profile and pioneering the application of ion beams for purposes beyond pure research. Under his guidance, GSI became the birthplace of ion-beam cancer therapy in Europe, treating its first 450 patients long before such treatments became standard clinical practice. He also played a pivotal role in establishing Europe's first Hadron Cancer Therapy Center (HIT) in Heidelberg, providing a bridge between nuclear physics and medicine in ways that have saved many lives. His far-reaching vision extended into plasma physics, leading to yet another transformative initiative during his tenure: the launch of the Petawatt High Energy Laser for Heavy Ion Experiments (PHELIX), designed to explore new frontiers in plasma physics. During Hans's tenure, I had the opportunity to visit GSI, where I delivered a colloquium, and I met him often at CERN. At CERN, he also served as Chair of the PSCC Committee and as a member of the Scientific Policy Committee (SPC).

Hans was a man of vision, an outstanding scientist—I would say a true force of nature—for whom nothing was impossible.

Carlo Rubbia
CERN
Geneva, Switzerland

Contents

1 **Introducing Hans Joachim Specht** 1
   Thomas Nilsson, Jürgen Debus, Shoji Nagamiya,
   and Dietrich von Harrach

2 **Legacy and Impact** 19
   Sanja Damjanovic, Volker Metag, and Jurgen Schukraft

3 **Hans Joachim Specht (in His Own Words)**
   Sixty Years of Physics—The
   Fascination of Diversity 27
   Hans Joachim Specht

4 **Reflections and Tributes** 135
   Peter Armbruster, Ewald Konecny, Volker Metag,
   Reinhold Schuch, Horst Schmidt-Böcking, Helmut Satz,
   Jurgen Schukraft, Michael Albrow, Axel Drees, Jochen Wambach,
   Wolfram Weise, Charles Gale, Dinesh Kumar Srivastava,
   Gianluca Usai, Ralf Rapp, Volker Koch, Hans Günter Dosch,
   Peter Schneider, Ulrich Charisius, Hartmut Eickhoff,
   Thomas Kühl, Andreas Tauschwitz, Klaus Dieter Gross,
   and Sanja Damjanovic

5 **Closing Thoughts** 227
   Sanja Damjanovic, Volker Metag, and Jurgen Schukraft

Acknowledgments 229

The Authors 231

Chapter 1
Introducing Hans Joachim Specht

Thomas Nilsson, Jürgen Debus, Shoji Nagamiya, and Dietrich von Harrach

The Past that Guides the Future: From GSI to FAIR

Thomas Nilsson

It is both a privilege and a great responsibility to assume the role of Scientific Managing Director of GSI and FAIR at this pivotal moment. As we look ahead to the exciting scientific opportunities that the FAIR project will bring, we should also acknowledge the vision and dedication of those who laid the foundation for GSI's long-term future. With deep respect and gratitude, we remember Professor Hans J. Specht, whose passing is a profound loss to the scientific community. As the former Scientific Managing Director of GSI, Hans played a defining role in shaping its long-term future.

Hans J. Specht was appointed Scientific Managing Director of GSI in 1992, at a time when the heavy-ion synchrotron SIS18, the storage ring ESR, and the fragment separator FRS had recently become operational. Personally, I vividly recall the time

T. Nilsson (✉)
Facility for Antiproton and Ion Research and GSI Helmholtz Centre for Heavy Ion Research (FAIR-GSI), Darmstadt, Germany
e-mail: t.nilsson@gsi.de

J. Debus
Heidelberg Ion Beam Therapy Center (HIT), Heidelberg, Germany
e-mail: juergen.debus@med.uni-heidelberg.de

S. Nagamiya
KEK, Tsukuba-shi, Japan

RIKEN, Wako-shi, Japan
e-mail: nagamiya@post.kek.jp

D. von Harrach
Institut für Kernphysik der Johannes Gutenberg Universität Mainz, Mainz, Germany
e-mail: harrach@uni-mainz.de

© The Author(s) 2025
H. J. Specht et al. (eds.), *Hans Joachim Specht*, Springer Biographies,
https://doi.org/10.1007/978-3-031-92353-1_1

when Hans first assumed the role of Scientific Managing Director. I was in the middle of my Ph.D. studies and a frequent visitor to the laboratory, where we had just completed the first complete kinematics experiment using radioactive beams from the FRS. The experimental campaign had been long and arduous, and its success had only been possible due to personal interventions and strong support from the previous Scientific Managing Director, Paul Kienle. Thus, my young colleagues and I anticipated the change in leadership and possible new strategies with mixed feelings, but our worries turned out to be completely unnecessary. Hans' leadership brought a fine crop of scientific results with the existing installations, while further opening up activities towards CERN and initiating crucial discussions and studies regarding the future expansion of the GSI facility during his tenure. This entailed significant involvement of the then Research Director Volker Metag, and broad participation from the European and international scientific community. Although the outcome of these studies and Hans' vision for GSI's future differed from the direction ultimately pursued, the strategic studies he led in the 1990s laid the groundwork for what would become the FAIR project—a facility which will continue to expand the boundaries of nuclear and particle physics for generations to come.

Under his leadership, characterized by thinking and acting "outside-the-box," GSI achieved internationally recognized milestones, particularly in the pioneering use of heavy-ion beams and lasers. One of his most significant achievements was the establishment of the pilot project for ion-beam cancer therapy. The first successful treatments of patients with C ions at GSI, sometimes to the mild inconvenience of other users who were confined to doing experiments at night, paved the way for the Heidelberg Ion Beam Therapy Center (HIT), Europe's first hadron therapy clinic. This groundbreaking achievement not only revolutionized the treatment of certain classes of cancer but also opened new frontiers in applied research, including plasma physics. The successful transfer of this experimental therapy to industry remains the paramount example of applications of GSI research in society and was important in securing the positive assessment of the FAIR project proposal.

Thanks to Hans' support and guidance, GSI furthermore played a pivotal role in the ultra-relativistic heavy-ion program at CERN from its inception. Beginning with the first heavy-ion experiments at the SPS in 1984 and culminating in the establishment of the ALICE experiment at the LHC, GSI's contributions have been instrumental in advancing our understanding of the early universe and the behavior of matter under extreme conditions.

Hans J. Specht's legacy is deeply embedded in the development of GSI and in pioneering studies and initiatives that paved the way for its long-term future. His unwavering commitment to scientific excellence, innovation, and international collaboration continues to inspire us all. As we advance toward the realization of the FAIR project, we do so with profound respect for his immense contribution.

Memories of Hans Joachim Specht

Jürgen Debus

I had my first encounter with Professor Specht as a physics student at Heidelberg University in the mid-1980s, where I attended his lectures on experimental physics. These lectures were of unprecedented intellectual intensity and also delivered at great speed. He held the chalk in his right hand and wrote on the black board, then wiped away his writing with the sponge in his left hand so as not to waste any time.

It was a shock when I was informed that he had been assigned as an examiner in my diploma examination. Although he was considered a highly objective examiner, he was feared because he enjoyed diving deep into countless many different areas of physics. My oral exam began with the harmonic oscillator, then led from the vibrating string to the potential well, followed by the whole story of charged particles, not forgetting relativistic effects, of course. This was followed by the hydrogen atom, the nuclear model leading to the strong interaction, and finally the Schrödinger equation. For sixty minutes I rode a roller-coaster through the world of physics. The atmosphere was friendly, however, and Professor Specht maintained an encouraging but piercing look throughout.

Five years later, in 1990, I met Hans Specht again at the inauguration ceremony of GSI's synchrotron, SIS18. At the time, he was on the GSI experimentation committee and I was a medical physicist and an attending physician collaborating closely with Gerhard Kraft's biophysics group, which was dedicated to researching radiation damage in tissue caused by charged particles. Professor Specht had an incredibly good memory. He remembered that I had taken my diploma exam with him and that it had gone rather well. Our Heidelberg delegation at the celebration consisted of Michael Wannenmacher the then director of the Heidelberg Radiology Clinic, Günther Gademann, the leading senior physician, and myself. We were particularly interested to hear about the progress on the planned therapy project at the accelerator there.

Before that, in the early 1980s, during the planning phase of the large particle accelerator SIS at GSI, the first contacts were established by GSI's director Gisbert zu Putlitz and Karl zum Winkel, former director of radiology at Heidelberg University. Building on the initial experiences of the clinical project at LBL Berkeley, California, plans were made for a clinical center with a dedicated accelerator for ion-beam therapy, including construction and investment cost planning. However, the funding bodies could not be convinced to finance the project as there was hardly any relevant work that supported the application. These relations were further strengthened by Gerhard Kraft, who headed the GSI biophysics group, and Michael Wannenmacher when he took over the radiology department in Heidelberg and simultaneously headed the experimental radiotherapy unit at the German Cancer Center (DKFZ) in Heidelberg under the directorship of Harald zur Hausen. In 1988, just before SIS18 became successfully operational, a new proposal titled "Construction of an experimental heavy-ion therapy at GSI Darmstadt" was submitted to the German government by the three cooperating institutes GSI, Heidelberg university hospital, and DKFZ.

Unfortunately, the ministry never acted upon the proposal. Nevertheless, developments at GSI continued with a relatively small amount of EU funding. The lesson from Berkeley was that a 3D dose distribution had to be generated, but the aim was to move away from the scattering and collimation process towards an active irradiation method. The dissertation by Thomas Haberer, who was developing the active raster scanning method, was funded as part of the EULIMA (European Light Ion Medical Accelerator) project. Financed by the European authorities in Brussels in cooperation with CERN, the EULIMA project brought together expertise from different laboratories to study accelerator options, beam delivery to the patient, and radiobiological properties of carbon ions compared to heavier ions. Importantly, socioeconomic studies of patient recruitment in different countries were carried out to demonstrate the need for several therapy units across the whole of Europe. When these first results came in, more and more people became convinced that a therapy project with carbon ions could be realized. Furthermore, the idea that GSI had already started working on a future Heidelberg therapy center could be more and more substantiated.

However, at that time, the vast majority of researchers at GSI did not want to be involved in the project, except for beam production. The great fear—another lesson learned from LBL—was that ion-beam therapy would be given absolute priority and lead to serious cuts in beam time for the nuclear and atomic physics programs at GSI.

This view changed completely in the spring of 1993 when Hans Specht became Director of GSI. He was willing to support the project with the highest priority in all technical, physical, and radiobiological details, in whatever areas GSI had the necessary experience. He wanted to see the first patient treated within the first 4 years of his directorship.

Immediately after taking office, he visited all the working groups and had them explain what they were working on. During this time, I became responsible for the clinical part of the project after Günther Gademann received an offer to go to Magdeburg. As Professor Specht and I knew each other already, there was immediately a basis of trust between us. I quickly noticed that if he was convinced of someone, they would have the greatest supporter one could imagine. He then paved the way for the success of the therapy project. For 2 years, whenever resources were needed, Specht provided them. The other physicists at GSI received less beam time. They had to accept this and Specht stood firm. Even when things got difficult, he believed in the project and removed obstacles. This was a remarkable achievement.

Independently of the preparation of a new proposal for the government, the construction of a therapy system began immediately in May 1993. The ambitious goal was to construct a heavy-ion therapy system based on active beam delivery by raster scanning, including fast energy variation by the accelerator, online quality control and in situ PET diagnosis, and single beam-spot treatment planning with local RBE assignment depending on the tissue and the radiation field. For the beam delivery system, a prototype of the raster scan system was already operating successfully, including prototypes of beam monitoring systems. But we had no previous experience of the new accelerator controls.

A major challenge was the fast beam control system, i.e., the monitors in front of the patient that guarantee the exact position and intensity of the particle fluence for each pixel. The beam monitoring system is one of the most critical components of the therapy unit because the speed of the control system is the limiting factor for the overall treatment time and hence for the number of patients that can be treated per year. At the beginning of the project, the readout of the multi-wire chambers (MWPCs) was still too slow, and the feedback mechanisms between the scanning system and the beam detectors also had to be developed. This was one of Thomas Haberer's tasks during his postdoc period, in addition to coordinating the development of the accelerator control system with Hartmut Eickhoff.

Moreover, novel solutions had to be found for beam quality control, treatment planning, and dose verification. It was obvious that an integrated project was required not only to develop new techniques in beam scanning, dosimetry, and treatment planning but also to demonstrate the clinical feasibility of patient treatments. Such a pilot project was proposed at GSI in 1993 by Gerhard Kraft and Dieter Böhne in cooperation with Heidelberg university hospital and DKFZ. Hans Specht set the course for the project to be funded. He was a brilliant speaker who knew how to present arguments and was able to inspire and convince in a perfectly factual way. Finally, the so-called "pilot project" was accepted, with the aim of starting patient treatment by the end of 1997.

The tight schedule of only 4 years to realize this novel technology was a great challenge. Dieter Böhne became general project leader. As head of the accelerator division, he had considerable experience in carrying out complex projects. The various sub-projects were coordinated by the committee KAT (Koordinations Ausschuss Therapie) chaired by Gerhard Kraft, while the whole project was supervised by an external Advisory Committee Therapy (ACT) chaired by Edward Alpen, LBNL, Berkeley.

A crucial point was the development of a suitable treatment planning system (TPS) that addressed the problem of dose adjustment for the delivery of scanned beams and included a model for the relative biological effectiveness (RBE). Since there was no commercial TPS for scanned ion beams, a new TPS was developed based on existing components developed at GSI and DKFZ. The research TPS called VIRTUOS (Virtual Radiotherapy System) developed at DKFZ was one of the first 3D planning systems in the world and provided the clinical functionality of image registration and segmentation and enabled the definition of treatment plan parameters and evaluation of dose distributions. This was combined with a dose engine TRP (Temporary Raster Planning) to calculate and optimize the carbon ion dose, developed by Michael Krämer at GSI and later renamed TRiP (Treatment Planning for Particles). This combination was used for clinical treatment planning at the DKFZ and GSI.

For dose prescription, the increased effectiveness of ions in relation to photons—which result from the Bragg peak—had to be taken into account in treatment planning.

The complex dependencies of the RBE on the dose, biological endpoint, position in the field, and so on, require a physical beam transport model covering all fragmentation processes and a very precise biophysical model. High accuracy is required since steep gradients are observed—and needed—in the dose response curves at the transition between tumor and healthy tissue. As a consequence, even small uncertainties in the estimation of the biologically effective dose can result in large uncertainties in the clinical outcome. Thomas Haberer developed the transport model YIELD that was later integrated into TRiP. Regarding RBE modeling, Michael Scholz did groundbreaking work at GSI in developing the Local Effect Model (LEM), which was continuously developed over the following years.

In a partnership with the Dresden-Rossendorf Center, a highly innovative quality control system was established under the leadership of Wolfgang Enghardt. For the first time, in situ PET measurements of the stopping beam distribution were performed on a routine clinical basis. At each fraction during the beam application, the PET camera measured the location of the stopping carbon isotopes. From this information, the quality of each fraction could be assessed and possible errors in the beam application could be corrected. The in situ PET control quickly became a central approach in this experimental therapy, because we were now able to determine the accuracy of the range distribution with an independent measurement after the end of each irradiation.

A separate treatment area for patients had to be built into the large accelerator hall. It was important to "hide" the technology behind panels as far as possible so that the patients were not frightened and felt as though they were in a clinic. It was essential for the success of the therapy project that all the different components for beam monitoring and application were integrated into a single control system that could be handled by the clinical crew. Our clinical team developed and practiced all clinical workflows with a dummy called Paula. When the final system was up and running, I decided to perform a "stress test." I opened the door to the treatment room during irradiation (of Paula) to check whether the beam really would stop and the interlock would be properly documented in the system. I measured the time we'd need to evacuate Paula from the immobilized treatment position in case of sickness or emergency. I also pushed a couple of wrong buttons. Of course, I did not tell the people in the control room before. Everything worked just as it should!

Shortly thereafter, just a few months before the first patient was to be irradiated, a recently appointed external technical director resigned unexpectedly. Specht had to decide whether to quickly find and appoint a successor or put the therapy project on hold for the time being. The situation was tricky. My colleagues at GSI and I, who had worked directly on the pilot project, were convinced that the project could be realized successfully and in time, but there was still the possibility that this departure could seriously delay the project. However, Hans Specht didn't take long to find a solution. He called Thomas Haberer and asked him if he would take over. He agreed and we were able to carry on.

Commissioning and QA protocols of the experimental carbon ion irradiation system, developed by the DKFZ team under the leadership of Günter Hartmann and later on by Oliver Jäkel, were completed on time. In December 1997, we obtained the approval of the authorities. We were ready for our first patient.

The plan was to treat only two patients, then monitor their progress over 6 months and carefully check for adverse effects. Two patients suffering from brain tumors were chosen, who were to receive a basic treatment with conventional radiotherapy in Heidelberg and a carbon ion boost of 5 fractions at GSI. A Saturday was chosen to start the treatment at GSI, because we did not want to have to deal with the routine experimental and administrative rush at GSI. No one but those absolutely necessary to run the system were allowed in the control room. Even Hans Specht waited in a nearby room respecting the need for calm and concentration. We prepared the patient carefully and then Thomas Haberer pressed the button for the first treatment beam.

Everything went well. All controls and measurements showed the expected values, and likewise for the second patient, who was scheduled that same day. Soon the space around the therapy cave filled with all the people who had worked so hard to make such treatment possible, and we celebrated this great success. The following set of five planned treatments also went well.

After these first two patients, there was an accelerator shutdown for 6 months which allowed us to see whether there were any unexpected effects, mainly because of the high local doses during beam scanning. Although we did not expect unforeseen adverse effects, we could hardly wait to see the clinical results.

Both patients were well. We were able to continue.

In the summer of 1998, the second treatment block, the clinical phase of the GSI pilot project, got under way. Treatments continued on a regular basis with three beam times of 4 weeks per year and 20–25 patients per beam time. One week was needed to set up the therapy mode parameters at the accelerator and 3 weeks to treat the patient. The rest of the time, the beam time belonged to the physics programs at GSI.

The patients were diagnosed with CT and MRI scans in Heidelberg and, if necessary, with PET, and then included in a clinical trial. Treatment planning began in the conventional way with target segmentation and the designation of possible entrance ports and organs of risk. This planning was done by the responsible physician and medical physicist in Heidelberg. These plans were then completed at GSI with the biological optimization programs. Once the plan had been approved by the medical physics team headed by Oliver Jäkel, the final plan was translated into steering files for the scanner system. The irradiation itself only took about twenty minutes, depending on the tumor size.

Patients were either shuttled daily 60 km by bus from Heidelberg or they stayed in a nearby hotel. Some patients used the time they were being treated at GSI as a vacation in the nearby recreation areas or continued working in the afternoon after the daily treatment. In one exceptional case, a patient came by bicycle a few hundred kilometers from Switzerland, stayed nearby during the treatment and, after 20 fractions, returned home again on his bicycle. These examples illustrate just how well patients tolerated the treatment.

By September 1998, we had demonstrated the clinical feasibility of carbon ion treatments with the raster scan technique and submitted a joint proposal to the Federal

Ministry of Research and Science for the funding of a clinical facility in Heidelberg. Hans Specht handed over the proposal personally to Minister Rütgers when he attended the official inauguration ceremony of the GSI pilot project.

In 1999, Hans Specht left GSI and handed the reins to Walter Henning, but our proposal was still under review by the German Science Council (Wissenschaftsrat der Bundesregierung) until 2001. Then, in May 2001, it was finally evaluated. Positively. Heidelberg University and the university hospital thus finalized the plans for the Heidelberg Ion beam Therapy center (HIT). At GSI, the department known as "accelerators for therapy" was founded under the leadership of Hartmut Eickhoff who, together with Thomas Haberer, was in charge of planning the accelerator system and the beam application system for the HIT.

In May 2003, the Science Council finally approved the plans and recommended the HIT project of the University Medical Center Heidelberg for funding with 50% of the total costs of 119 million €. Construction work began that same year.

Meanwhile, we went on treating patients at GSI. Up until 2008, 434 patients were treated in over 10 years of clinical operation. The main indications were skull base chordoma and chondrosarcoma (treated typically with a 60–66 Gy RBE weighted dose in 20 fractions), adenoidcystic carcinoma, sacral chordoma, and advanced prostate carcinoma (all treated typically with a boost of 18 Gy RBE weighted dose in 6 fractions after conventional radiotherapy). All treatments were optimized with LEM I using an α/β-ratio of 2 Gy. Clinical trials for chordoma, chondrosarcoma, and adenoid carcinoma showed excellent tumor control rates, and in the case of adenoid carcinomas, these were higher than in a comparable photon collective. Patient follow-up continued, even after the end of the project, resulting in 5y- and 10y-local control rates of 72%/54% for chordoma, 88%/88% for chondrosarcoma, and 59.6%/42.2% for adenoidcystic carcinoma, respectively. As a result of these clinical trials, the safety and effectiveness of carbon ion treatments were demonstrated, and chordoma, chondrosarcoma, and adenoidcystic carcinoma were established as standard indications for carbon ion therapy in Germany, and reimbursed by many health insurances.

In 2009, HIT began clinical operations and of course Hans Specht was present at the inauguration ceremony. By this time, Hans Specht had become professor emeritus and I was chair of Heidelberg radiation oncology and had become CEO and Medical Director of the HIT, while Thomas Haber had become Technical Director of the HIT. Hans Specht told us that day what a moving experience it was for him to see our shared idea realized in a clinical center with three levels, fully integrated into the oncological services of the Heidelberg campus, including the university hospital, the DKFZ, and the National Center for Tumor diseases (NCT).

It is no exaggeration to say that, without Hans Specht, there would have been no ion-beam therapy at the GSI, hence no HIT in Heidelberg. His commitment to this medical innovation is a legacy that will benefit many generations to come.

We continued to meet regularly after his "retirement" which was, of course, action-packed, and although our conversations no longer had an exam-like feel to them, they were always intense. We never had small talk.

Hans Specht was an outstanding physicist, but he also had broad scientific interests and was ready to discuss a wide range of biological and medical topics. He was open to other opinions as long as they could be backed up by data and logical thinking.

I will conclude with some of his own thoughts, shared with us in his farewell speech at the GSI in 1999. He said that, in order to successfully create new things, we need clear responsibilities coupled with the willingness to lead. He himself exemplified this like no other. In his speech, he also emphasized that success in science is based on individuals who are given the means to work creatively on new things in small teams, and then allowed to make suggestions and follow them through. He was convinced that successful innovation requires a clear willingness to take personal responsibility and to take risks. He received thunderous applause. Even though this speech was given 25 years ago, it has lost none of its relevance. At the end of this speech, he talked of his finest hour at GSI. He was visibly moved when he described the happiness in everyone's eyes after successfully irradiating the first patient as the best moment of his career. He concluded with the words: "that's when we understood why we do science."

May Hans Specht rest in peace. Let us take his example as an inspiration and carry forward his passion for science, truth, and life in our own actions.

My Recollections of Hans J. Specht

Shoji Nagamiya

I met with Hans J. Specht for the first time in Heidelberg in May 1982 when he decided to change his field of research from low-energy nuclear physics to high-energy heavy-ion collisions. At that time, a workshop on Quark Matter Formation and Heavy-Ion Collisions was held in Bielefeld, Germany, where he and I became co-organizers of a session called *Inclusive Measurements and Particle Identification*. According to the records, Hans must have been 45 years old at the time, and was based at the University of Heidelberg, while I was 37 years old and employed at the University of Tokyo, having just returned from Berkeley, where I had worked at the Bevalac.

Hans was extremely friendly right from the start, always smiling. I also vividly remember him proudly showing me his very speedy car, which was almost completely made of plastic. We drove through the middle of Heidelberg at a thrilling speed of 180 km/h. Hans even offered to drive me to Bielefeld at an even higher speed, but I was filled with fear and excitement and politely declined his offer.

Hans then moved to CERN in 1983, taking a sabbatical leave, and started the SPS heavy-ion experiments. His most famous contribution came later from the CERES experiment, which led to the discovery of an enhanced production of low-mass e^+e^- pairs. This was a puzzle for a long time, but we later found out that it was due to the decay of the ρ meson into a continuum of lepton pairs due to the partial formation of the quark–gluon plasma. This discovery significantly advanced our understanding of heavy-ion collisions and the behavior of the quark–gluon plasma, a key concept in high-energy heavy-ion physics. This work will be described in further detail later in this book.

After our first encounter in 1982, we became very close friends and often discussed a wide range of topics together. When Hans became the Scientific Managing Director of GSI in Darmstadt in 1992, he asked me to serve as a member of the "Wissenschaftlicher Beirat" (Scientific Advisory Board), a role I held until the end of his term. In those days, immediately after the completion of the heavy-ion synchrotron SIS18, the use of SIS heavy-ion beams came under discussion. On Hans's initiative, we debated their use in nuclear physics and other fields, including medical applications.

There were two significant events in 1996, 15 years after I first met Hans Specht. The first was the international conference Quark Matter'96 in Heidelberg in May 1996, where Hans served as the chair. I, of course, attended this conference since I have always enjoyed visiting Heidelberg. The second event was Hans's 60th anniversary celebration, held at GSI, which I also attended. The year 60 is a special occasion in Europe, but it is also significant in China and Japan since the Chinese calendar recycles every 60 years. Two photos remain (Fig. 1.1), which I got from Sanja Damjanovic. In each, I have enlarged Hans Specht, while cutting the surrounding image.

In 2007, we organized a major international conference on nuclear physics in Tokyo. As the chair of the conference, I recall Hans Specht asking me whether he really needed to attend. Finally, he decided to come to Japan. It was indeed a wonderful conference since both the Emperor and Empress came to talk with selected audience members. Of course, Hans chatted with the Emperor and enjoyed it very much. This was after Hans had retired from the university.

Throughout 2010s, I often visited GSI. On 19 December 2012, I was invited to give a talk there and met with Hans Specht once again. He was 76 at that time. This was more than 30 years after I first met him. Many heavy-ion physicists were

Fig. 1.1 Two encounters with Hans in 1996. Left: QM'96 in Heidelberg (Hans J. Specht, Helmut Satz, and Shoji Nagamiya, from left to right). © Hans J. Specht. All rights reserved. Right: The 60th anniversary of Hans Specht at GSI, with Shoji Nagamiya as the keynote speaker. © FAIR-GSI, with a photo credit to A. Zschau. All rights reserved

gathered there, including GSI's father of relativistic heavy-ion physics, Rudolf Bock, who passed away on 19 April 2024, shortly before Hans himself.

We all met with great sadness the news that Professor Hans J. Specht passed away on 20 May 2024, at the age of 87. His departure was a loss to the entire heavy-ion world, and it will leave a void for all of us. We sincerely hope his profound legacy will continue to inspire generations and ensure that he will live on even after passing from the Earth.

"Specht Here!"

Dietrich von Harrach

Specht here!
These were the familiar opening words of phone calls made on weekends or after work hours—words that, unsurprisingly, rarely delighted the wives of Hans Specht's team members. Such calls carried an unmistakable message: their husbands' immediate presence was required, whether in the detector lab, the measurement room, or at the computer for urgent data analysis. How was it possible that these employees, time and again, abandoned their wives, children, and weekends to embrace these pressing demands so eagerly, without hesitation, complaint, or excuse?

The answer to this question can only be found by delving into the concepts of "charisma" and "charismatic personality." The term, originally derived from the realm of religion, refers to the natural authority of an enlightened figure. This person speaks the truths of the universal spirit and, through the recognition of this, inspires followers—disciples, so to speak. Since it seems impossible to learn or fake charisma, the truly fascinating question is: How does a charismatic personality emerge—one that can set ambitious goals, illuminate paths to achieve them, and, in addition, gather loyal followers around him?

Since I do not consider myself capable of answering this question in any systematic way, I will instead attempt to approach it by recounting the life journey of Hans Specht. I feel compelled to do so, as I have counted myself among his "disciples" for more than 50 years. Furthermore, I am authorized to write about him, as Hans Specht himself specifically requested a contribution from me for his Festkolloquium on the occasion of his 80th birthday. Even then, the central theme of my reflections was the question of charisma.

Hans Specht came from a bourgeois background, born into a family of furniture manufacturers who valued higher education, and particularly the fine arts. Despite his early talent for playing the piano, he chose his second passion—natural science and technology—as his professional calling. Intelligence and curiosity drove him to delve deep into the heart of things, constantly seeking a deeper understanding. He turned to academia and left the management of the kitchen furniture factory, near the "Kamener Kreuz," to his brother.

For reasons I can only speculate on, he chose Munich for his physics studies—a city that, in the 1960s, was one of the most exciting centers of post-war German physics. The first German reactor in Garching, with Heinz Maier-Leibnitz as its founder, the fresh fame of Rudolf Mössbauer, and the towering figure of Werner Heisenberg made Munich a place of both cutting-edge excellence and intellectual continuity. Hans worked on his diploma and doctoral thesis under the guidance of Peter Armbruster at the Garching reactor, focusing on the detection of atomic excitations in fission products. It was here, in this exceptional environment of scientific innovation, that his academic journey took root and began to flourish. These works illuminate two recurring themes: the dynamic processes involved in nuclear fission and the development of novel detection methods for ions. Building gas counters with ultra-thin windows proved to be a challenging task, one that could not be solved by merely filling out an order form.

As a postdoc, Hans Specht spent some time at Chalk River in Canada to study the dynamics of the fission process. Upon his return to the Munich Accelerator Laboratory, he gained his first international recognition with the detection of rotational transitions in the strongly deformed second minimum of fission isomers. This was also the period when his initiative, experimental skill, and leadership clearly became evident. It was during this "mission (almost) impossible" that he gathered around him a group of enthusiastic and devoted doctoral students who spent sleepless nights waiting for the "ping" from the second minimum. Yes, it was there, and yes, the method was right. I still vividly remember one of his first students, Dietrich Habs, who would speak in awe of "Hacky's" remarkable talents.

What was it that made him so fascinating to young people? First and foremost, it was his remarkable clarity—his ability to formulate goals and the means to achieve them with precision. Hans Specht referred to this approach as "straightforward thinking." To think straight and act straight requires a deep understanding of the terrain, the ability to avoid pitfalls and to overcome seemingly insurmountable obstacles.

Hans Specht was deeply fascinated by cutting-edge technology—photography, film cameras (see Fig. 1.2), spy tape recorders, sports cars, and Steinway pianos. I knew well his passion for brands like Olympus, Beaulieu, Bolex, Nagra, Lotus, and more. He delved into every detail, exploring them down to their very core. Where were the limits? Hans loved technology and admired people who knew just a little more or could do something extra—piano tuners, car mechanics, computer freaks—real professionals. He wanted to understand every detail and would "squeeze" them for every last bit of information. It could be exhausting, but it was also immensely helpful—being "pushed" to your own limits and then perhaps, together, finding a new question to explore.

He was especially attentive when it came to matters of electronics and the emerging world of online computers. I myself, due to certain circumstances, became a user and programmer of a PDP-8, and I could do something that caught his interest. Despite my excellent dissertation, they didn't want me in Munich—I was considered too radical. But Hans Specht, freshly appointed as a professor in Heidelberg, couldn't have cared less. He called a postdoc position in Copenhagen "a waste of time" and

1 Introducing Hans Joachim Specht

Fig. 1.2 Hans Specht had a lifelong passion for photographic cameras, movie cameras, and lenses. An avid collector, he acquired a wide variety of cameras spanning several decades of technological advancement and featuring iconic brands such as Olympus, Beaulieu, Bolex, Leica, Zeiss Jena, Kern, and P. Angénieux, among others. Notably, his impressive collection includes a significant number of cameras from the French company Beaulieu, renowned for its expertise in super 8 and 16 mm film cameras. Apart from collecting, he made movies during family vacations, using the different camera types and more often than not, trying the patience of all involved with his perfectionism during the process, even though everybody was happy with the final films and photos in the end. © Hans J. Specht. All rights reserved

said that an assistant position in Heidelberg, with the newly established GSI nearby, was exactly the right thing for me.

It was, and it established a relationship of trust with very wide latitude. The first joint project was the development and construction of the so-called "Großmaulzähler" (large-mouthed counters) for the "Käseglocke" (cheese dome) at GSI, designed to study heavy-ion-induced fission reactions. Uranium on uranium, above the Coulomb barrier, was a "sexy" topic, promising the jackpot of discovering superheavy elements. However, it also involved mastering the bread-and-butter events with four fission products. This is where I learned from Hans Specht the method of kinematic coincidences. Today, it might seem like a triviality, but back then, it was a groundbreaking innovation: large solid angles with many particles in the final state. The proponents of the classical method, who were used to shuffling magnet spectrographs or tiny semiconductor counters around—pixel and strip counters were still dreams at that time—coined the term "Großmaulzähler" as a somewhat playful jab at the new approach.

Of course, Hans Specht was always one step ahead, and some colleagues felt uneasy that, in every discussion, in committees, or in seminar rooms, he always had the final word—cool and sharp-witted. The art of writing a successful grant proposal could be learned from him—always thinking and writing straight to the point.

I must absolutely tell you about the birth of the "Großmaulzähler": the massive scattering chamber at the Z4 beamline at GSI. The methods for reconstructing high-energy scattering events, beyond the capabilities of photographic plates, were developing rapidly at the time, particularly through the groundbreaking work of Heidelberg's high-energy physicists Heintze, Soergel and Walenta. The gas counters for minimally ionizing particles could "easily" be set up in a hall because energy loss in the air wasn't a significant factor. However, the enormous ionization densities of the heavy ions or fission products required a vacuum and ultra-thin entrance foils for the low-pressure gas counters. The challenges of building a setup with enough flexibility to accommodate vacuum modules were overwhelming. But not for Hans Specht, who simply declared, "Then we'll just pump the hall out." No sooner said than done: the "Käseglocke" was born. Not the entire hall, but pretty close.

New Physics from 1980

The physics of heavy ions at energies just above the Coulomb barrier—referred to by Paul Kienle as "clump physics"—is particularly captivating. In these processes, the dynamics is primarily driven by the high angular momenta of the clusters, which, during scattering, largely preserve their identity, apart from diffusive nucleon exchange processes. What makes this so fascinating is the delicate interplay between macroscopic and microscopic degrees of freedom, reminiscent of fission dynamics. These interactions can be effectively captured using Fokker–Planck transport equations.

It is "too hot" to isolate the low-energy collective excitations necessary for "cold" fusion—an approach pioneered by Peter Armbruster and his colleagues, which led to their glorious successes in synthesizing new heavy elements. Yet, at the same time, it is "too cold" to observe subnuclear excitations, starting with pion production. Even at these energies, central collisions do not result in significant density changes, meaning that the "quarks" must remain invisible—though everyone knows they are there.

The rapid successes of the quark–parton model and quantum chromodynamics (QCD), driven by experiments at CERN and DESY—where Heidelberg physicists played a pivotal role—were at the center of scientific attention. In contrast, the "clumps" had relatively little to offer. The real action was undoubtedly taking place in Geneva or Hamburg—or in American laboratories—where you could "see" quarks and gluons, provided you learned the language of structure functions and jet fragmentation.

It was at this point that a seminar project was born, organized by Hans Specht together with Jörn Knoll for interested nuclear physicists, held in a small back room at the Max Planck Institute on Saupfercheckweg. We learned at breakneck speed about scaling behavior, hadronization, and Drell–Yan processes through review articles by leading CERN physicists, enabling us to grasp the revolutionary nature of the first α–α "heavy-ion" experiments at the ISR storage rings at CERN. For all of us involved, this marked a pivotal turning point in our scientific orientation, shifting our focus towards the relationship between QCD and nuclei.

A lecture by Bill Willis, the driving force behind the ISR ion experiments, at GSI, followed by a reciprocal visit from Specht's team to CERN, was all it took to trigger this phase transition. Hans Specht then joined Willis' group at CERN during a sabbatical, and subsequently began developing proposals for his own experimental program in new collaborations at the heavy-ion beam of the PS/SPS. GSI already played a pivotal role here, as it was able to provide the ion sources. Our paths diverged at that point, when I joined the group of Bogdan Povh, whose focus was on deep inelastic scattering on nuclei and the newly discovered EMC effect.

GSI Director

At that time, GSI had already been involved in relativistic heavy-ion programs at Berkeley and had achieved a moderate energy increase through the SIS, enabling experiments beyond the meson thresholds. As director, Paul Kienle had been a driving force behind the development of storage rings for highly charged heavy ions. When a successor was sought who could bring new impetus to GSI, especially in the direction of subnuclear degrees of freedom, Hans Specht was the obvious candidate. However, he found himself in a dilemma, as the natural goal of conducting experiments at the highest possible baryon densities was already achievable at CERN with their existing accelerators and the ion sources developed with GSI's help. A race to catch up at GSI by building accelerators in the range of several tens of GeV/n would be difficult to justify and could only be defended through unique selling points such as luminosity or precision.

I have the impression that this situation persists to this day, as the pressing questions regarding the equation of state of matter at high densities—such as those related to neutron stars and black holes—are still on the table, and may perhaps be answerable at the LHC. Meanwhile, the race to catch up at GSI seems hindered by a lack of resources and personnel. However, one must acknowledge that heavy-ion physics at CERN essentially benefits "for free" from the ongoing research into fundamental interactions—W, Z, top quarks, Higgs, and SUSY.

As an alternative to ultra-relativistic heavy-ion collisions, the so-called heavy-ion inertial fusion could have been a viable option. At that time, this approach was in direct competition with the laser inertial fusion strategy pursued in the United States. To this day, I have never fully understood why this option, or even a combination of the two as drivers and diagnostics, was not pursued further or considered worth pursuing.

In this complex situation, Hans Specht turned to the strengths of GSI in the production of precisely controllable heavy-ion beams. GSI had long since established a department dedicated to studying the biological processes in tissue cells under heavy-ion irradiation. The controllable range and ionization density were already well known, as were the raster scanning techniques used for tumor treatment. To develop this into a clinical application, close collaboration between radiation therapists and accelerator and radiation physicists was essential. However, the scientific cultures of medical professionals and physicists are, by necessity, quite different. Only a few truly understand how to bridge these differences.

A remarkable example of a researcher who could connect these cultures and mediate between them was the Heidelberg physician and physicist Hermann von Helmholtz, who notably combined the physical and physiological effects of acoustics and hearing. Anyone who attended the Friday colloquium at Philosophenweg 12 would have been familiar with the display case showcasing the original Helmholtz resonators in the foyer.

I am confident that Helmholtz's legacy played a pivotal role in inspiring the transformation of radiation therapy, which, in close collaboration with exceptional medical professionals from Heidelberg, was turned around and set on the right course. I've already mentioned the exhausting interviews that Hans Specht conducted with the experts he recognized and deeply admired. Many of these interviews were crucial in identifying the obstacles—technical, medical, and political—that needed to be overcome, all through the process of "straightforward thinking." I've already touched on the persuasive power and charisma that Specht brought to this task.

In short, it worked, and the breakthrough was lasting. Additional tumor centers were established, utilizing the methodology that had been developed. With that, GSI was once again freed from the compromises it had to make with the other culture, enabling it to turn its focus to new and exciting objectives.

The development of tumor therapy, together with the groundbreaking creation of superheavy elements, not only elevated but firmly cemented GSI's national and international reputation.

This is not meant to downplay the successful research in the fields of spectroscopy and nuclear reactions, which today plays a significant role in advancing our understanding of element formation in the universe. However, on a more critical note, GSI spent an unusually long time explaining some of its "discoveries." There seems to have been a lack of "straightforward thinking" to avoid such "time-wasting." There was no shortage of charismatic figures, though.

Hans Specht focused on the future direction of GSI with deep commitment, thoroughly investigating every possible avenue. We had already discussed the dilemma posed by the existence of the CERN accelerators, along with their antiproton facilities and the radioactive beams they generated.

The study of a future program in working groups also included the idea of an Electron–Ion Collider, a research direction aimed at exploring the strongly interacting substructure of bound nucleon systems at normal nuclear densities. Nonlinear effects, such as so-called shadowing or the EMC effect, also play a role in the study of nucleus–nucleus collisions. It was no coincidence, then, that Hans Specht appointed me as one of the coordinators. A storage system was designed, and the advantages of studying the hadronic final state were clearly outlined. The use of spin degrees of freedom in momentum transfers, where the application of perturbative methods from quantum field theory make sense, was also considered. Methods for increasing luminosity were discussed. The concept of electron–nucleon collisions was not new and, in principle, had potential in the HERA storage ring system in Hamburg, as demonstrated in experiments, at least for polarized electrons and unpolarized deuterons.

Here, GSI ventured into treacherous terrain—the domain of Bjørn Wiik, the undisputed driving force behind German high-energy physics in the 1990s. HERA was his creation, and the next step was the proposed electron–positron linear collider. I am certain that this project would have been realized had it not been for his untimely death in an accident. In any case, Wiik held all the reins and sought to align every project—at least within Germany—with this vision. Any deviation towards electron–nucleon collisions would, at best, have been regarded as merely "riding along" in the wake of CERN's heavy-ion program. Wiik, however, would have gladly welcomed the new community, regardless of where it came from. I can speak from experience, having been involved in the relevant working groups under Bjørn Wiik and his successors, and having served as a reporter on numerous occasions.

It soon became clear what happened to the GSI plans: they melted away like butter in the sun under Bjørn Wiik's charm offensive. Here, Hans Specht had found a worthy rival—but one who held the better cards.

It is a bitter irony of history that the showdown between the large-scale projects TESLA and FAIR, before the Science Council and under the successors of Wiik and Specht, ended in favor of a mixed concept–one without a clear vision or well-defined goals. As a side note, the EIC project, an Electron–Ion Collider, has now become one of the few major projects dedicated to hadron structure physics, naturally utilizing spin degrees of freedom. Out of respect, Andreas Schäfer and I were invited to the first meeting of the founding initiative at MIT, but that was the full extent of our involvement.

Today, DESY and GSI, at best, play a secondary role in addressing the fundamental questions of matter and the cosmos, primarily through their participation in experiments at CERN. However, the X-ray laser and the research into the stability and frequency of nuclei—classic topics in atomic and nuclear physics—should not be downplayed. It may be that this reflects the financial constraints faced by our country, but it is also possible that the challenge lies with the available personnel. Originality, courage, and strength are essential for identifying and answering the crucial questions of nature. Hans Specht never lacked in any of these qualities—originality, courage, or strength.

As a final thought, I would like to reflect on Hans Specht's profound and lifelong love for piano and chamber music—a passion that deeply inspired and guided him throughout every aspect of his life. It brings to mind the spirit of Beethoven's late string quartets, compositions imbued with an unparalleled sense of urgency, originality, and intensity—works that his contemporaries found overwhelming. Legend has it that Beethoven inscribed these stirring words on the score: *"Muss es sein?—Es muss Sein!"* ("Must it be?—It must be!").

Open Access This chapter is licensed under the terms of the Creative Commons Attribution 4.0 International License (http://creativecommons.org/licenses/by/4.0/), which permits use, sharing, adaptation, distribution and reproduction in any medium or format, as long as you give appropriate credit to the original author(s) and the source, provide a link to the Creative Commons license and indicate if changes were made.

The images or other third party material in this chapter are included in the chapter's Creative Commons license, unless indicated otherwise in a credit line to the material. If material is not included in the chapter's Creative Commons license and your intended use is not permitted by statutory regulation or exceeds the permitted use, you will need to obtain permission directly from the copyright holder.

Chapter 2
Legacy and Impact

Sanja Damjanovic, Volker Metag, and Jurgen Schukraft

Hans Joachim Specht, one of the founders of ultra-relativistic heavy-ion physics and a pioneering figure in hadron cancer therapy, passed away on 20 May 2024, at the age of 87. A graduate of the Ludwig Maximilian University of Munich (LMU), the Technical University of Munich (TUM), and ETH Zurich, and full professor at the University of Heidelberg for more than 30 years, his career was distinguished by important contributions across a whole spectrum of scientific domains.

Early Contributions to Physics—Shaping Atomic and Nuclear Physics

Hans J. Specht started his academic career in atomic and nuclear physics in Munich, under the guidance of Heinz Maier-Leibnitz.

Opening the Field of Quasi-Atoms

Hans J. Specht's pioneering research in atomic physics, particularly during his Ph.D., laid the groundwork for the study of quasi-atoms, opening new frontiers in heavy-ion atomic physics. His discovery of resonance-like peaks in the ionization cross sections of inner electron shells observed during near-adiabatic collisions of heavy ions—first detected through X-ray production in fission products—marked the beginning of quasi-molecule research. In his seminal 1965 paper in *Zeitschrift für Physik*, Hans J. Specht provided the first qualitative explanation of this phenomenon in terms of quasi-molecular orbits, introducing the transformative concept of quasi-molecular states as key drivers of electron transitions during collisions. Notably,

---

S. Damjanovic (✉)
GSI Helmholtz Centre for Heavy Ion Research, Darmstadt, Germany
e-mail: sdamjano@cern.ch

V. Metag
II. Physikalisches Institut, Justus-Liebig-Universität Giessen, Giessen, Germany
e-mail: volker.metag@exp2.physik.uni-giessen.de

J. Schukraft
CERN, Geneva, Switzerland
e-mail: jurgen.schukraft@cern.ch

© The Author(s) 2025
H. J. Specht et al. (eds.), *Hans Joachim Specht*, Springer Biographies,
https://doi.org/10.1007/978-3-031-92353-1_2

these insights emerged in an era with only a very limited theoretical framework. Today, this work is recognized as the discovery of the "level-matching effect." His pioneering interpretations catalyzed decades of research on heavy-ion collisions and X-ray production mechanisms, cementing his legacy as a visionary in atomic physics. Today, superheavy quasi-molecules are seeing a revival with the new heavy-ion machines, thus carrying Hans J. Specht's original idea to another level.

Discovery of Shape Isomerism—A Landmark in Nuclear Fission

Another highlight of Hans J. Specht's career was the discovery and precise measurement of shape isomerism in heavy nuclei, a phenomenon which brought new insights into nuclear fission. Hans J. Specht's observation of distinct rotational bands in plutonium-240 showed for the first time that nuclei can be in a highly deformed cigar-shaped state shortly before fission, confirming the concept of a "double-humped" fission barrier due to shell effects at large deformations. The parameters of the barrier were determined using transmission resonance spectroscopy, a groundbreaking approach, and a pivotal experiment in the field at that time. Later in Heidelberg, by measuring the quadrupole moment with the innovative charge-plunger technique, the strong deformation of fission isomeric states could be determined quantitatively to correspond to an axis ratio of 2:1. This discovery was considered one of the most important advancements in the shell model since the introduction of the spin–orbit coupling.

In Munich, and also later in Heidelberg, Hans J. Specht developed several innovative large-scale detectors for fission fragments and the reaction products of heavy-ion collisions, becoming one of the leading experimentalists in the new field of heavy-ion physics, with experiments at the Max Planck Institute for Nuclear Physics (MPIK) in Heidelberg and at the newly founded "Gesellschaft für Schwerionenforschung" (GSI) in Darmstadt (now the GSI Helmholtz Center for Heavy-Ion Research).

Pioneering Ultra-Relativistic Heavy-Ion Physics at CERN

In the early 1980s, Hans J. Specht reoriented his research towards the higher energies available at CERN, marking a new stage in his career. His contributions and advocacy, alongside a handful of other enthusiastic proponents, were instrumental in establishing CERN's ultra-relativistic heavy-ion program at the Super Proton Synchrotron (SPS) accelerator, which was approved in 1984, enabling studies of nuclear matter under the extreme conditions which existed in the early Universe. Achieving this groundbreaking frontier required innovative detector concepts and advanced experimental techniques.

Hans J. Specht became the spokesperson of a first-generation heavy-ion experiment (HELIOS/NA34-2), initiator and spokesperson of a second-generation experiment (CERES/NA45), a leading force in the analysis of the NA60 experiment, and a crucial supporter of the third-generation ALICE experiment at the Large Hadron Collider (LHC).

Hans J. Specht was a brilliant experimentalist with a keen eye for cutting-edge detector concepts and how to apply them with a minimalistic approach. This was apparent in his masterpiece, the dilepton experiment CERES, which used a "hadron-blind" double Cherenkov detector and a specially crafted magnetic field configuration. These innovations enabled the experiment to pick out and measure the properties of those rare electrons—prominent probes which preserve the full memory of nuclear collision—in the haystack of hadrons. The groundbreaking success of CERES, despite initial skepticism, earned it the affectionate nickname "the Spechtometer"—a tribute to Hans J. Specht's vision and determination in proving the feasibility of what many thought was impossible. Hans J. Specht often quoted a memorable remark from his esteemed colleague Sam Ting: *"If you want to make a major discovery, build a dilepton detector."*

The Quark–Gluon Plasma Journey—Discoveries through Dileptons

Initially with CERES, and later as a leading force within the next-generation NA60 experiment at CERN's SPS, Hans J. Specht succeeded in detecting, for the first time, thermally produced lepton pairs in heavy-ion collisions. The original discovery with CERES/NA45 remains one of the most cited papers from the SPS heavy-ion program. The high-precision measurements in NA60 of one of the most challenging signals, the Planck-like spectrum of thermal radiation at higher masses, and the precise characterization of the in-medium modification of the ρ-meson at lower masses, proved to be crucial in establishing the existence and properties of quark–gluon plasma (QGP)—the state of strongly interacting matter thought to have existed in the primordial Universe, just a few microseconds after the Big Bang. The enduring quality and relevance of these measurements, taken under his guidance, remain unsurpassed almost two decades later.

Contributions to International Science Policy

Throughout his career, Hans J. Specht held numerous prestigious positions in the realm of science policy at a variety of German and international research institutions. At CERN, he notably served as chair of the Proton Synchrotron and Synchro-Cyclotron Committee (PSCC) and as a member of the Scientific Policy Committee (SPC). He was also a founding member of the first board of directors of the European Centre for Theoretical Studies in Nuclear Physics and Related Areas (ECT*) in Trento, a place which held special personal and professional significance for him. As a member of the Nuclear Physics European Collaboration Committee (NuPECC), he contributed significantly to fostering collaboration and advancing nuclear physics research across Europe.

Visionary Leadership and Pioneer in Ion-Beam Cancer Therapy

Hans J. Specht also provided fresh impetus for nuclear and particle physics research as Scientific Managing Director of GSI (1992–1999), where he set the technical and science-policy course for the development and application of a groundbreaking innovation in radiation medicine—ion-beam cancer therapy. A pilot project at GSI for the irradiation of tumors with carbon-12 ions, launched under his leadership, successfully treated 450 patients and led to the establishment of the Heidelberg Ion-Beam Therapy Center (HIT), the first European ion-beam therapy facility. Reflecting on his achievements, Hans J. Specht was most proud of his contributions to ion-beam therapy, which advanced human health through the application of cutting-edge science and underscored the societal impact of fundamental research. Beyond this, Hans J. Specht initiated discussions on the long-term future of GSI, which eventually led to the proposal for the International Facility for Antiproton and Ion Research (FAIR).

Hans J. Specht also played a key role in inspiring the development of the proton therapy center in Trento through his close interactions with Renzo Leonardi. Both ECT* and the Trentino region owe him a great deal. Hans J. Specht and Ugo Amaldi were the main figures behind the initiation of the 5th Hadron Cancer Therapy project in Europe—the South East European International Institute for Sustainable Technologies (SEEIIST) Project.

Exploring the Intersection of Physics, Music, and Neuroscience

Hans J. Specht's curiosity extended beyond physics. He also had a profound interest in the overlaps between physics, music, and neuroscience, collaborating with Hans Günter Dosch on understanding the perception of music and its physiological bases. Their joint lecture, "Physics and Music," first presented in 1986 on the occasion of the 600th anniversary of the founding of the University of Heidelberg, developed into an independent research program, which initially focused on psychoacoustics, and then expanded into neurophysiology through collaboration with the Neurological University Hospital in Heidelberg. This transdisciplinary approach led to surprising discoveries in auditory processing and produced highly cited publications (including *Nature Neuroscience*) on the differences in the auditory cortex between musicians and non-musicians, pushing the boundaries of how we understand the brain and its response to music. Hans J. Specht's global influence in this field is underscored by numerous lectures worldwide delivered jointly with Hans Günter Dosch, including the Loeb Lectures at Harvard in 1999 and the prestigious 4th Einstein Lecture Dahlem in 2006.

A Lifelong Passion for Mentorship and Teaching

Hans J. Specht was an outstanding teacher, a prolific mentor, and a successful science manager, but first and foremost, he was an inquisitive scientist—someone who had a profound love for physics, a thirst for knowledge, a hunger for understanding, and a relentless drive to follow wherever his interests and research would lead him.

Hans's lectures in Heidelberg were unforgettable for many of his students. In 2001, he was honored as the *Best Lecturer for the Winter Semester 2000/2001* in Physics III—Relativity and Quantum Physics, based on student evaluations conducted by a special committee. Motivated by this recognition, Hans created a unique set of lecture notes titled "Experimental Physics 3—Introduction to Modern Physics: Relativity, Quantum Physics and the Structure of Matter." Praised for their clarity, distinctive style, and use of vivid, real-world examples, these notes continue to be highly regarded by students, reflecting Hans's exceptional talent for teaching physics, and his ability to make complex concepts both accessible and inspiring.

A Life Beyond Science—Passion for Cars

Hans J. Specht's frequent and spirited commutes between Heidelberg and CERN in his iconic green Lotus Elan over more than 40 years will be fondly remembered. His love for cars was well known among colleagues and friends, and his passion for speed ran deep (Fig. 2.1).

Remarkably, his very first publication in 1960 was not about physics but about tuning his car for greater speed. At just 24 years old, Hans J. Specht authored the article *ABARTH-UMBAU*, published in *mot Roller Mobil Kleinwagen* (Issue 7/1960)—from the same publisher as *Auto Motor und Sport*. Titled *"Mehr PS für den Jagst"* (*"More Horsepower for the Jagst"*), the piece detailed his work on tuning a Fiat 600 Jagst for improved performance. Even decades later, Hans J. Specht spoke of this achievement with pride—a testament to his lifelong drive for optimization, whether in experimental physics, detector technology, or fine-tuning the perfect engine (Fig. 2.2).

An early photograph of Hans J. Specht with his family captures his fascination with classic cars—his three children, happily enjoying the outdoors and the thrill of the ride, in an era before seatbelts and modern safety features became the norm (Fig. 2.3).

As a professor at the University of Heidelberg, Hans J. Specht had the unique opportunity to further indulge his passion for motorsports by participating in the opening of a Formula 1 race in Hockenheim with his Lotus.

Fig. 2.1 Hans J. Specht's 55-year-old Lotus Elan, a faithful companion throughout his journey in physics. © Hans J. Specht. All rights reserved

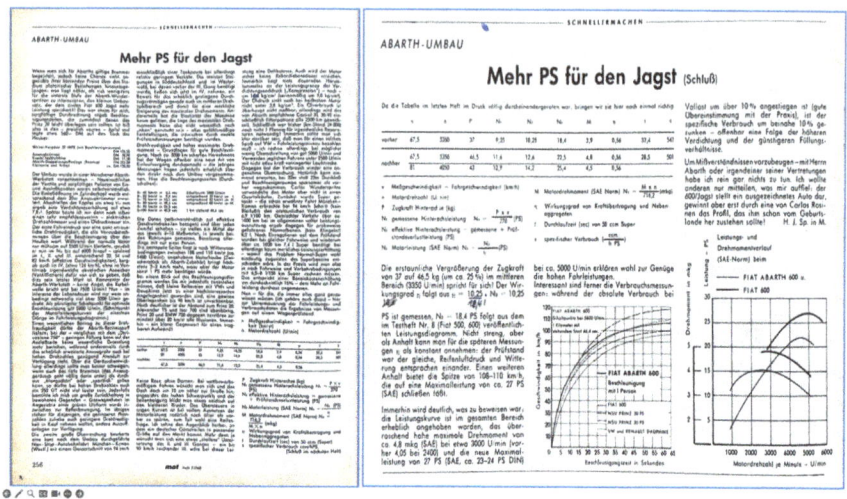

Fig. 2.2 Hans J. Specht's first publication—the article ABARTH-UMBAU, published in mot Roller Mobil Kleinwagen (Issue 7/1960), from the same publisher as Auto Motor und Sport, titled 'Mehr PS für den Jagst' ('More Horsepower for the Jagst'). © mot Roller Mobil Kleinwagen/Motor Presse Stuttgart GmbH & Co. KG/2025. All rights reserved

Fig. 2.3 Hans's idea of outdoor sports included—apart from swimming, sailing, and the occasional scenic hike—cruising with his Lotus sports car. With his family—his wife Adelheid and their three children Martin, Katja, and Michael—he often drove to Lake Starnberg near Munich for swimming. He especially enjoyed driving at high speed on Alpine pass roads and even took part in the 24-h race at the Nürburgring racetrack. Figure 1971, Munich. © Hans J. Specht. All rights reserved

Hans J. Specht was a luminary in the scientific community, whose unwavering curiosity and pioneering contributions continue to inspire generations of scientists. His legacy is a testament to the enduring power of curiosity-driven research and visionary leadership—a life devoted to the joyful pursuit of scientific discovery and the betterment of society.

His critical guidance and profound questions will be deeply missed by all who had the privilege of knowing him.

Open Access This chapter is licensed under the terms of the Creative Commons Attribution 4.0 International License (http://creativecommons.org/licenses/by/4.0/), which permits use, sharing, adaptation, distribution and reproduction in any medium or format, as long as you give appropriate credit to the original author(s) and the source, provide a link to the Creative Commons license and indicate if changes were made.

The images or other third party material in this chapter are included in the chapter's Creative Commons license, unless indicated otherwise in a credit line to the material. If material is not included in the chapter's Creative Commons license and your intended use is not permitted by statutory regulation or exceeds the permitted use, you will need to obtain permission directly from the copyright holder.

Chapter 3
Hans Joachim Specht (in His Own Words) Sixty Years of Physics— The Fascination of Diversity

Hans Joachim Specht

This memoir is the written version of a lecture I delivered in Heidelberg on 28 January 2016, as part of the series "*Heidelberg Physicists Report: Emeriti Reflect on Their Scientific Journeys.*" During the lecture, I presented over 60 slides, featuring particularly intriguing physics results alongside their corresponding experimental setups, as well as many photographs of my mentors, students, and other important encounters in my life. A video recording of this lecture also exists, and it would be impossible to capture the full breadth of that content in print. Therefore, this written version should be seen more as a complement to the lecture than as a faithful reproduction, incorporating many additional notes and reflections which were not part of the slides or oral presentation. Despite the effort required to meticulously prepare long-forgotten and poorly documented material, I found the lecture quite enjoyable. It was important for me to express my gratitude to the many mentors from whom I learned, as well as to the countless students who initially learned from me, but who have in their turn given far more back to me.

- *Lecture given on 28 January 2016, in Heidelberg*
  Link to the slides:
  https://www.physi.uni-heidelberg.de/~specht/material/slides/Emeritus_Talk_HD-Jan2016.pdf
  Link to the video:
  https://www.youtube.com/watch?v=rKtfuCb0r8Y&list=PL6xB_3xN4Acv8WVhHVneBXvjIvRL4KmDg&index=6

H. J. Specht (✉)
Physikalisches Institut der Universität Heidelberg, Heidelberg, Germany

From My Childhood and Early Passion for Science to a Lifelong Career in Physics

My Youth

Born in 1936 in a small village in Westphalia, I grew up relatively undisturbed by the turmoil of the war and the immediate post-war period, surrounded by a rural environment. My years at the Gymnasium in Kamen did little to alter this provincial life. How I came to be interested in physics remains somewhat unclear. My family background was shaped by a company my grandfather founded in 1905, which continues to thrive into its fourth generation today. Among my siblings and numerous relatives in my parents' generation, there were medical doctors, lawyers, and even teachers in the previous generation, but no scientists. During my youth, I was interested in many things: astronomy using telescopes I made myself, electronics by building radios and remote controls, and even chemistry (conducting experiments which, in hindsight, were quite risky). Some of these interests had early roots: at the end of the war in 1945, we children, including some older than me, collected baskets full of electronic materials, including field telephones, from a destroyed army depot nearby. In 1946, we also collected gunpowder from US ammunition after the local command had vacated our house. Later, I eagerly read all kinds of books about physics, including atomic and nuclear physics, but also technical applications of physics.

Regarding my love for physics and music, I was passionate about both, but music was the primary focus of my time. In my parents' generation, in keeping with the bourgeois tradition in the first half of the twentieth century, almost everyone played an instrument, usually the piano, as I did. In high school, my particular strengths were mathematics, physics, and music. As I was not shy about speaking up and had a tendency to dissent (there was also dissent in the family against the Nazi regime, led by my grandfather), I was elected class representative for the last 3 years. I was also top of the class.

Academic Beginnings in Munich—Early Years 1956–1965

After graduating from high school in 1956, I enrolled at the Ludwig Maximilian University in Munich (LMU) to study physics, feeling that physics offered the best unifying framework for all my interests. Any remaining doubts completely vanished during that first semester, which I view more as a general study experience due to my regular attendance of lectures from other faculties, particularly in the medical field. I have never regretted this choice: the fascination with both the fundamental and the richly application-oriented aspects of the field continues to captivate me to this day. At the same time, Munich offered an intellectually, socially, and also (admittedly) touristically stimulating environment, which had a profound impact on young students from the provinces.

Diploma (1962) and Doctoral Dissertation (1964) in Atomic Physics at FRM Munich

I obtained both with summa cum laude, completing the doctoral thesis in under 2 years. The details can be summed up rather quickly. Initially, I transferred to the Technical University in Munich, motivated by the less abstract mathematics (under J. Lense) and, in particular, by the broad range of supplementary subjects beyond experimental physics (under G. Joos), such as descriptive geometry, technical thermodynamics, and especially technical electrical engineering (under W.O. Schumann). For the mandatory 6-month industrial internships, I spent half of the time in a repair workshop for steam locomotives, where I also learned essential metalworking skills, and the other half in the chemical laboratory of a medium-sized steelworks in the Ruhr area, owned by a member of the family.

After completing my preliminary diploma in 1959, I moved to ETH Zürich but struggled with high-level theoretical physics, as Wolfgang Pauli, whose lecture notes were circulating in Munich, passed away unexpectedly before the start of the first semester in Zürich. Instead, I focused on advanced experimental physics lectures, particularly on high-frequency technology with a substantial practical component, which later became my chosen subject for the main diploma (under H.H. Meinke). Strongly influenced by the 1959 Lindau Nobel Laureate meeting, which featured M. Born, P. Dirac, O. Hahn, W. Heisenberg, M. von Laue, W. Lamb, and other prominent figures, as well as Munich colleagues who shared exciting developments in modern physics from Garching, I returned to the Technical University in Munich. There, I completed my diploma thesis (Diploma 1962) and doctoral dissertation (Ph.D. 1964), both summa cum laude, under H. Maier-Leibnitz at the first German nuclear reactor "Forschungsreaktor München" (FRM). This was followed by an additional year as a postdoctoral researcher in Munich (Fig. 3.1).

Fig. 3.1 The early years (1956–1965). Studies at LMU Munich, the Munich Technical University, and ETH Zurich (1956–1959); Diploma in 1962; Ph.D. in 1964. The photo marks the completion of the Ph.D. © Hans J. Specht. All rights reserved

Reflections on My Early Years

In retrospect, what most profoundly shaped my scientific development up to the beginning of my CERN years was, I believe, the outstanding personality of H. Maier-Leibnitz and the exceptional environment he cultivated in Munich.

Regarding nuclear physics, the spirit of the times was far more positive than it is today. In 1956, the agreement for the purchase of the Garching nuclear reactor ("Atomei") was signed by the Minister of Atomic Affairs, Franz Josef Strauß, in the presence of H. Maier-Leibnitz. Construction began that same year, and, astonishingly, the reactor became operational just 11 months later—unthinkable compared to today's overregulated environment. At that time, fundamental research in this field was still in the making, with many open questions, both qualitative and quantitative. The use of neutrons for structural analysis in solid-state physics, chemistry, and biology was still in its infancy, accompanied by a constant stream of new ideas and high expectations for the future (Fig. 3.2).

In the realm of technical applications, despite the dark shadows cast by Hiroshima, Nagasaki, and the global arms race, there was also tremendous enthusiasm for a broad range of more constructive applications following the first major Geneva Conference on the "Peaceful Uses of Atomic Energy" in 1955. Students from around the world, including myself from ETH, were drawn to this emerging field. At any given time, there were "simultaneously 100 diploma students and 100 doctoral students," all under the supervision of H. Maier-Leibnitz and a handful of assistants, guided by the principles of "everyone is responsible for their work" and "everyone helps everyone else"—original quotes from Maier-Leibnitz's published Emeritus Lecture in Heidelberg, 1992.

I learned two key lessons. Firstly, the profound importance of developing new methods to continually gain fresh insights in experimental sciences, as summed up by Maier-Leibnitz's philosophy, inspired by Lichtenberg: "do something new to see something new." This principle has influenced my own scientific methodology to this

Fig. 3.2 Contract signed for FRM in 1956. © Hans J. Specht. All rights reserved

day. Secondly, I learned the value of teamwork in both directions: learning from the more experienced as a young diploma student and, as a doctoral candidate, mentoring four younger Ph.D. students during the final 6 months of my own relatively short one-and-a-half-year project. Such collaborative structures were crucial for the survival and success of Maier-Leibnitz's system.

Efforts to radically improve the situation were indeed successful, for in 1963, at the initiative of Maier-Leibnitz and his colleagues, and supported by his former student, R. Mössbauer (Nobel Prize 1961) who was keen to return to Germany, a departmental structure was approved. The new department began its work in 1965, just before I left Munich. The number of chairs increased from 9 to 16, and a total of 240 positions were created.

Postdoc in Chalk River/Canada 1965–1968

I then undertook a postdoctoral position at the Chalk River Nuclear Laboratories in Canada from 1965 to 1968, at the invitation of J.C.D. Milton and J.S. Fraser, based on an NRC Fellowship, which was likely arranged by Maier-Leibnitz. Prominent figures such as A. Bromley and T. Litherland, who had already moved on to other positions, made frequent visits. Other postdocs there included D. Pelte and O. Häusser. The latter was a highly talented cellist with whom I often performed music (even publicly, though only in that remote, culturally-isolated setting). A personal highlight was a 1-week visit from H. Maier-Leibnitz during this period. In my daily work and during the frequent beam times at the EN and MP tandem accelerators, I was largely left to my own purposes. The universities were too far away to involve students, and my wonderful senior supervisors were overloaded with pioneering work for the world's first spallation neutron source, the Canadian ING project, which, unfortunately, was never approved.

Professor at LMU Munich 1969–1973

In any case, three harsh winters in Canada were enough for me. Despite the efforts of my hosts to keep me there, I returned to Munich in 1969, this time to the Ludwig Maximilian University (LMU), which offered promising development opportunities in the newly established Physics Section run by S. Skorka. Together with his colleagues J. de Boer, U. Meyer-Berkhout, and C. Zupancic, Skorka supported my habilitation in 1970, based on my Canadian research. This was followed by my appointment as an HS2/3 professor in 1971. My significant role in establishing the joint Munich Accelerator Laboratory renewed old connections with the Technical University of Munich (TUM), particularly with P. Kienle and E. Konecny. Despite the apparent rivalry between LMU and TUM, I felt at home in both environments, which provided me with unforeseen advantages. My subsequent research successes quickly gained recognition, including in Heidelberg, where I first made my mark by being invited to deliver a colloquium in the summer semester of 1972.

Professor at Heidelberg University: Lifelong Impact and Commitment

Next, I spent a guest semester in Heidelberg from 1972 to 1973, where I was entrusted with delivering the main introductory physics lecture, a tradition established by O. Haxel. The invitation came from Herr zu Putlitz, whom I barely knew at the time, in a truly unforgettable phone call. Shortly thereafter, in 1973, I received several offers and accepted the position at the (then II.) Physics Institute in Heidelberg, succeeding O. Haxel. Although I felt a touch of nostalgia for the beloved Bavarian environment, I was thrilled with the new opportunity. The subsequent merger of the two institutes into a department-like structure, with colleagues such as J. Heintze, G. zu Putlitz, and V. Soergel, created a unique atmosphere of harmonious communication and collaboration, which extended into our personal lives. This supportive environment was further enhanced by the complete freedom to choose my research focus and my enjoyment of teaching: despite some gaps, I taught 5 out of 6 core experimental lectures in rotation (and was consistently rewarded with excellent diploma students). My dedication to Heidelberg remained steadfast despite several attractive offers to move elsewhere. In 1983, I declined an official offer to Mainz to lead the establishment of MAMI. A few years later, I also halted semi-official discussions regarding a position at Columbia University after a visit and a colloquium, to avoid disappointing my esteemed colleagues T.D. Lee (Nobel laureate) and W.J. Willis.

A New Period from 1983—High Energy Heavy-Ion Physics at CERN and Scientific Managing Director of GSI

Throughout these years, this dedication was reflected in my three sabbaticals at CERN from 1983 to 2003 and a 7-year secondment to serve as the Scientific Managing Director of GSI Darmstadt from 1992 to 1999.

Research at the FRM and Accelerators 1961–1983

My scientific interests up until around 1983 fall into three core research areas, all interconnected by the common theme of low-energy physics, in contrast to high-energy physics:

- **Atomic physics**, focusing on inner electron shells (which was also the subject of my dissertation).
- **Nuclear physics**, with a primary focus on nuclear fission, a topic which deeply captivated my interest and engaged my research throughout this period.
- **Heavy-ion reactions**, which opened up an exciting new frontier with the advent of the first uranium beams at GSI's UNILAC facility in 1976.

Atomic Physics

The World's First Accelerator for "Heavy Ions"

The first area of research, atomic physics, was established somewhat fortuitously at the FRM in Munich. P. Armbruster supervised my diploma thesis, and I was likely the first person he mentored even before completing his own doctorate. Our relationship, characterized by deep scientific respect and genuine personal warmth, began to develop within the first few months of our collaboration and has endured to this day—almost unique longevity in my professional life. At the reactor, Armbruster meticulously constructed a gas-filled mass separator using two magnets left over from the CERN PS in 1959. In this setup, fission fragments produced by neutron-induced fission in a ^{235}U target near the reactor core were subsequently transported through a vacuum tube and separated with a mass resolution of approximately 4%. This setup was the world's first "accelerator" for genuinely "heavy ions," based on a prototype developed at Oak Ridge National Laboratory, Tennessee. Acceleration was achieved through the Coulomb repulsion between the two fission fragments, reaching energies of 1.0 and 0.5 MeV/u for mass numbers <A> equal to 100 and 140, respectively, representing the two groups of the asymmetric mass distribution. Although the setup was a groundbreaking achievement, its intensity—just 300 per second—was orders of magnitude too low for nuclear reactions, although sufficient for atomic reactions.

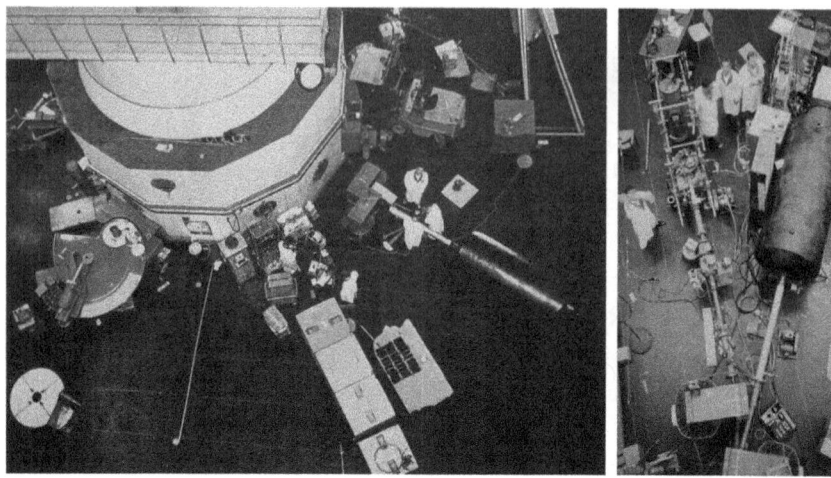

Fig. 3.3 Left: Beam line for the 'mass spectrometer' at the FRM reactor in Munich (photos from 1959/1963–1964). This setup was part of the 'atomic physics' topic addressed in Hans J. Specht's diploma and Ph.D. dissertations. Right: Hans J. Specht, together with Peter Armbruster and other Ph.D. students. © Hans J. Specht. All rights reserved

Consequently, the research program focused predominantly on nuclear spectroscopy of these highly radioactive fission fragments, which were captured on a thin foil. This focus led to several dissertations.

The Impact of Early Detector Development—Shaping My Career

During my diploma thesis, I was responsible for developing extremely thin window foils for the separator and for designing and building methane-filled transmission proportional counters with multiple wires. These counters were used to detect both β-radiation in coincidence with plastic scintillation detectors and passing fission fragments, the latter operating at pressures as low as 0.3 Torr. This work resulted in one of my first publications and proved beneficial for the research of other Ph.D. students. My time in Munich led to seven publications, including, to my great satisfaction, a contribution to the first IAEA conference on the "Physics and Chemistry of Fission" in Salzburg in 1965. This paper was co-authored by H. Maier-Leibnitz, P. Armbruster, and myself. My early involvement with detector developments profoundly shaped my entire subsequent career as an experimental physicist (Figs. 3.3, 3.4 and 3.5).

Discovery of "Quasi-Atoms"

My doctoral dissertation focused on the spectroscopy of characteristic X-rays emitted in atomic collisions of heavy ions, following the ionization of inner electron shells.

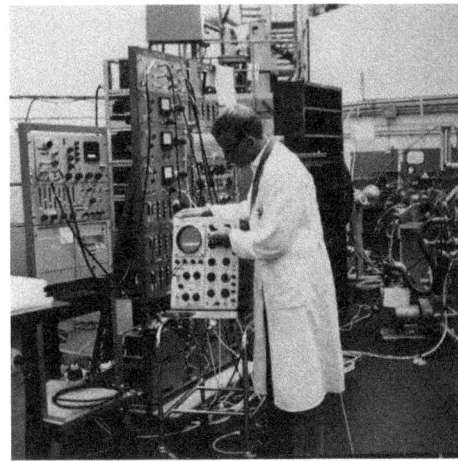 

Fig. 3.4 Hans J. Specht (left), and with Peter Armbruster and other Ph.D. students (right). © Hans J. Specht. All rights reserved

Fig. 3.5 Peter Armbruster (far left) and Hans J. Specht (far right). Peter Armbruster's habilitation, 1964. © Hans J. Specht. All rights reserved

P. Armbruster had previously described initial results in his dissertation, indicating unusually high cross-sections. My task was to investigate this phenomenon systematically by measuring cross sections in collisions with fission fragments across the full periodic system. Using a self-built, 1/2-m-sized counter filled with Ar/methane, I measured X-ray spectra from collisions of light and heavy fission fragment groups (<Z> values 38 and 54) on approximately 20 targets, ranging from Be to Pb ($Z = 4$–82). From these measurements, I determined the ionization cross-sections for the K-, L-, and M-shells separately for both collision partners. These results led to an unexpected discovery: a dramatic increase in cross-sections whenever the binding energies of electrons in certain inner shells (such as L/K, L/L, or L/M) coincided, indicating energy degeneracy of the corresponding states. At the time of my doctoral examination in 1964 (P. Armbruster had already left, having accepted a position in Jülich), this finding remained a great mystery, despite numerous discussions, including with theorists.

These discussions occupied a significant part of my doctoral examination. However, it was only months later, while drafting the publication and carefully studying G. Herzberg's well known textbook on molecular physics in Ottawa, that a breakthrough finally came.

The collisions were quasi-adiabatic, meaning that the nuclear motion was slow compared to the speed of the electrons in the inner shells. This quasi-adiabatic nature allowed the collision partners to be viewed as quasi-static molecules, with molecular states corresponding to the textbook description, extending to the limit of atomic states of the combined "quasi-atom" with nuclear charge $Z1 + Z2$ (Fig. 3.6).

The observed maxima in the ionization cross-sections could then be explained by hole transfer at the intersections of the corresponding states. This first interpretation in terms of quasi-molecular orbits, along with correlation diagrams and molecular

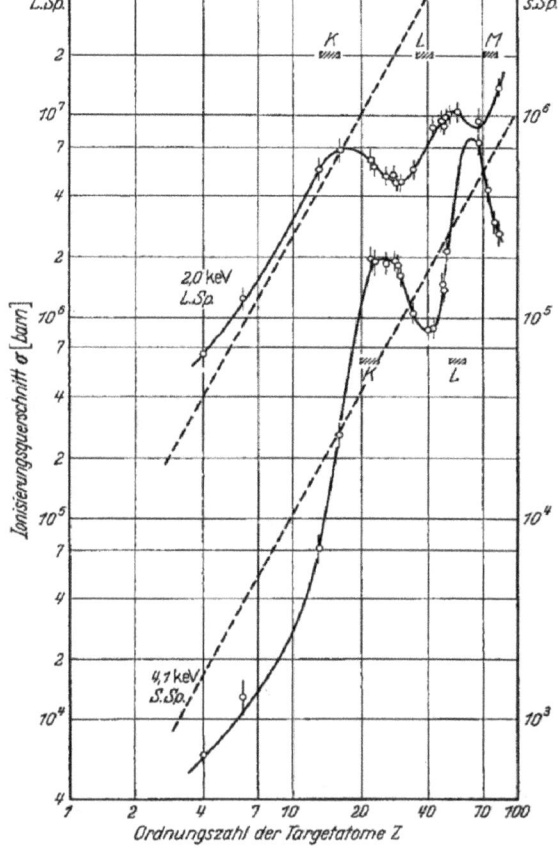

Fig. 3.6 Ionization cross-section of the L-shell of light ($<Z> = 38$) and heavy ($<Z> = 54$) fission fragments versus the Z value of the target atoms. Dashed line: Born approximation [Z. Physik 185 (1965) 301]. © 1969 Springer-Verlag. All rights reserved

orbital (MO) diagrams, is presented exclusively in the published version of my dissertation. This interpretation was submitted 2 months before the analogous, though more extensive, theoretical work by Fano and Lichten in *Physical Review Letters,* 1965. Their study, which was inspired by data for lighter ions such as Ar–Ar in the keV/u energy range, did not incorporate the observed Z-dependencies (Fig. 3.7).

Unfortunately, H. Maier-Leibnitz wanted the publication to be submitted to *Zeitschrift für Physik*, a German-language journal. As a result, my work was only discovered and cited for the first time in early 1970 by R. Brand et al. in *Physical Review Letters,* likely because the Western scientific community had long overlooked German-language journals. Today, this work is recognized as the discovery of the "level-matching effect." However, at the time, I felt quite disheartened by this delayed recognition, despite the eventual acknowledgment and correct interpretation of my results. I decided never to publish in German again. Joachim Heintze, in his 1992 emeritus lecture video in Heidelberg, reflected on a similar experience from 1957, when his fresh data on parity violation was published (with Nobel laureate J. Hans D. Jensen as editor), describing the "*Zeitschrift für Physik*" as "a first-class burial ground."

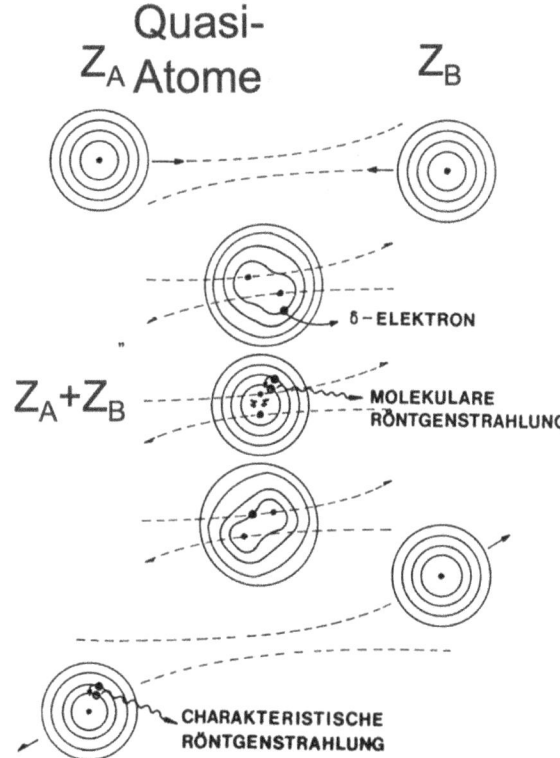

Fig. 3.7 Inception of the field of 'quasi-atoms'. © Hans J. Specht. All rights reserved

These results made a key contribution to starting a new field, undetected until 1969 due to the problem just described. It was only in the early 1970s that significant activity in this field began to develop at various accelerators, such as with P. Armbruster's group at the Cologne Tandem accelerator and, from 1976, at GSI's UNILAC facility. A few years later, R. Schuch joined me in Heidelberg, and together with the team, we published numerous papers between 1976 and 1984. Our team included many students, as well as external collaborators—notably H. Schmidt-Böcking from Frankfurt and occasionally I. Tserruya from the Weizmann Institute in Rehovot. The measurements were carried out at the MP Tandem Accelerator of the MPIK in Heidelberg. The focus was on investigating the impact parameter dependence, which benefited from the use of position-sensitive parallel-plate gas detectors, as well as on measuring molecular X-rays emitted during collisions when the two colliding partners closely approached in a quasi-atomic state. These efforts significantly advanced our understanding of collision dynamics, laying the groundwork for future studies in atomic and nuclear physics.

One of the most intriguing aspects of this research was the formation of quasi-atoms with atomic numbers of 184 (U+U) and the associated detection of the predicted "collapse of the neutral vacuum" through the emission of positrons in the "supercritical" electric fields, as described by W. Greiner et al. Despite extensive efforts over many years by groups including those led by E. Kankeleit, P. Kienle, and D. Schwalm at GSI, this search has not yet been brought to fruition because the collision times are too short and as a result the widths too large. Over the past 10 years, there has been a revival of the field in the investigation of metastable inner-shell molecular states or MIMS, i.e., the same states as before, now distinguished by the emission of molecular X-rays, but generated under extreme compression, such as in planetary cores or the interior of stars.

Selected Publications—A Guide

Early Detector Development

- Messung der Gasverstärkung in Methan bei niedrigen Drucken mit einem Transmissions-Proportional-Zählrohr für Spaltprodukte

  H.J. Specht and P. Armbruster

  Nukleonik 7 (1965) 8–14

  Expanded part of my diploma thesis, starting a life-long dedication to detector development. Lowest ever pressure in wire counters (0.3 Torr).

Atomic Physics: Inner-Shell Ionization in Heavy-Ion Collisions

- Ionisation innerer Elektronenschalen bei fast-adiabatischen Stößen schwerer Ionen

  H.J. Specht; Z. Phys. 185 (1965) 301–330

Expanded part of my Ph.D. thesis. Cross sections in collisions with fission fragments across the full periodic table, showing very large oscillations (detection of the "level matching" effect). First interpretation in terms of quasi-molecular orbits via correlation diagrams. A key contribution to starting a new field, undetected until 1969 due to the publication being in German. First hints of unexpectedly large cross-sections by P. Armbruster.

– Study of Impact-Parameter Dependent K-Vacancy Probabilities in Near Symmetric Gas and Solid Target Collision Systems

R. Schuch, R. Hoffmann, K. Müller, E. Pflanz, H. Schmidt-Böcking, and H.J. Specht

Z. Phys. A316 (1984) 5–14

End of a series of papers with R. Schuch et al. on characteristic and non-characteristic X-ray emission in heavy-ion collisions (started in 1976).

Nuclear Fission

The second area of research is nuclear physics, with a specific focus on nuclear fission. Discovered in 1938, nuclear fission, when examined through advancements in nuclear structure from the 1950s and 1960s, proved to be a highly complex process. With my publications in this field up to 1965, I was considered by my hosts, D. Milton and J. Fraser in Canada (from 1965 onward), as an expert in nuclear fission, while they were less interested in atomic physics. They themselves were regarded as one of the leading groups worldwide for their pioneering work on nuclear fission at their reactor, but felt it was time to move to more modern tools, such as the Chalk River EN Tandem Accelerator (Fig. 3.8).

Initiated by A. Bromley et al. via the High Voltage Engeneering Company (HVEC) and in operation since 1959 as the world's first, it had come to dominate international nuclear physics. Its seventh model was installed at the MPI for Nuclear Physics in Heidelberg in 1962. As a newcomer, my role was to develop the application of this precision technology to measurements of nuclear fission, a timely and promising endeavor in 1965.

After initial experiments on the (d,pf) reaction on the target nuclei ^{235}U and ^{239}Pu, using state-of-the-art silicon detectors for protons and fission fragments, we were fortunate enough to make a crucial theoretical breakthrough overnight. V. Strutinsky in Kiev had been working for years on a generalized shell model for atomic nuclei combined with the liquid drop model. According to his work, shell corrections lead to a second minimum in the fission barrier at an exact 2:1 axis ratio of the elongated nucleus. This minimum, driven solely by the neutron component, corresponds to an exceptionally strong binding, similar to the "magic" nucleon numbers associated with spherical nuclei, as interpreted by M. Göppert-Mayer and H. Jensen in Heidelberg in 1949. As illustrated on the right in Fig. 3.9 (excluding spin–orbit coupling),

Fig. 3.8 Chalk River Nuclear Laboratories (CRNL), Canada. © Hans J. Specht. All rights reserved

the nucleon numbers corresponding to gaps in the level scheme for spherical nuclei must be replaced by entirely different magic numbers when the nuclear shape is 2:1, and again by different numbers for a 3:1 ratio, with each configuration exhibiting distinct stability. The mysterious "spontaneously fissioning isomers," discovered by S. Polikanov et al. in Dubna in 1962, thus found for the first time a plausible explanation for their short half-lives—ranging from just 10^{-9} to 10^{-3} s: these isomers exist in the ground state of the second minimum, in contrast to the vastly longer half-lives of 10^4–10^9 years observed for spontaneous fission from the ground state of the first minimum. The "magic" neutron number for the 2:1 deformation in this context was later determined experimentally to be $N = 146$.

Pioneering High Resolution Nuclear Spectroscopy with a Novel Spark Chamber System

High-resolution nuclear spectroscopy in the second minimum became a priority and captured my attention, along with other projects, well into the 1970s. Silicon detectors, due to their insufficient energy resolution and limited surface areas, were quickly ruled out. However, the laboratory was equipped with a high-resolution Brown–Buechner magnetic spectrograph, similar to those in most tandem labs, including Heidelberg. Despite its potential, only photographic plates were being used in the focal plane. This was where my training in Munich came into play and proved crucial.

The final setup consisted exclusively of gas detectors and was developed practically single-handedly. The fission reaction under study was again ^{239}Pu(d,pf). The

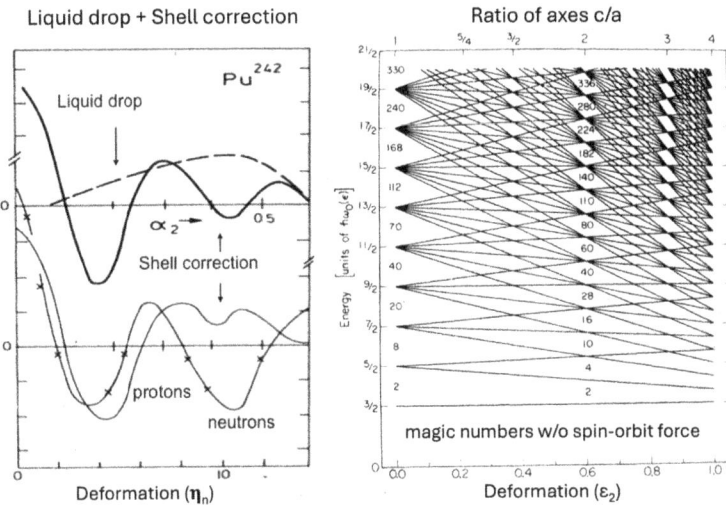

Fig. 3.9 Contributions of the liquid drop model and shell corrections to the fission barrier (Strutinsky 1966, left). Generalized shell structure within a harmonic oscillator potential (right). © 1974 American Physical Society. All rights reserved

magnetic spectrograph focused the protons along the focal plane. Position determination was achieved using a one-meter-long, hyperbolically shaped wire spark chamber, designed to match the shape of the focal plane. This was complemented by magnetostrictive readout. A 1-week stay at the BNL High Energy Physics department provided the necessary insights for implementing this design. The spark chamber was triggered by a fast parallel-plate proportional counter positioned behind it, operating in coincidence with a transmission proportional counter at the entrance of the spectrograph for particle identification through time-of-flight measurements. The fission fragments were detected in a cylindrical "multi-wire chamber" with 14 wire proportional counters surrounding the target, these being used to determine the spins of excited states from their angular distribution. This innovative replacement of traditional photographic plates in the new setup remained unique in nuclear physics for many years, even at MPIK in Heidelberg. The setup achieved a spectral energy resolution of 7 keV rms (Figs. 3.10, 3.11 and 3.12).

As a notable anecdote, the wires of the fission fragment chamber were, as was standard practice in Munich, separated by thin conductive intermediate walls to ensure precise potential ratios—a challenging task given the delicate dimensions. What I failed to realize, and now consider the greatest oversight of my professional life, was that shortly thereafter, in 1968, G. Charpak removed these walls in such detectors without encountering any significant drawbacks. This innovation paved the way for large numbers of wires and expansive wire planes—a true revolution in high-energy physics which earned him the Nobel Prize in 1992. H. Maier-Leibnitz's laconic comment at the time was: "We could have thought of that too."

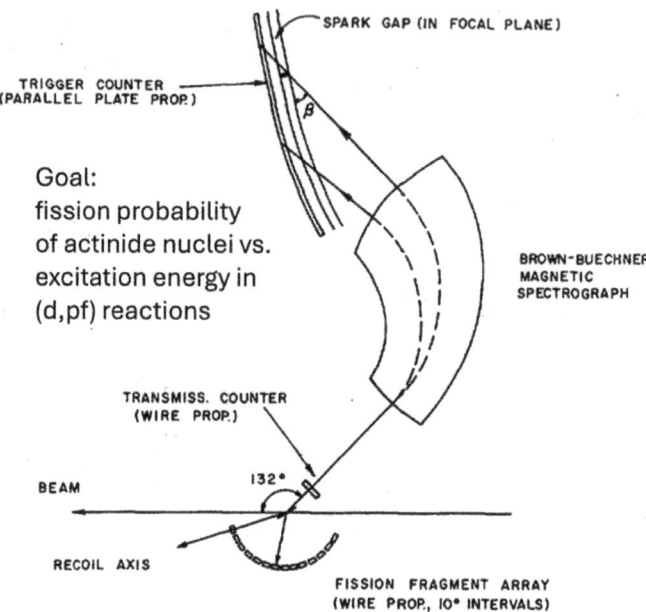

Fig. 3.10 A schematic drawing of the experimental setup at the EN Tandem in Chalk River. © IAEA 1969. All rights reserved. (IAEA publication: INTERNATIONAL ATOMIC ENERGY AGENCY, SPECHT, H.J., FRÄSER, J.S., MILTON, J.C.D., DA VIES, W.G., A high-resolution study of the 239Pu (d,pf)-reaction, Second IAEA Symposium on Physics and Chemistry of Fission, Proceedings of a Symposium Held in Vienna, 28 July–1 August 1969, IAEA,Vienna (1969) 363–375)

What was sensational at this time was that the fission barrier might be double-humped. The first transmission-resonance spectroscopy of the double-humped fission barrier was indeed made possible by replacing the then-common photo emulsions in magnetic spectrographs with a spark chamber (which was part of my habilitation thesis) (Fig. 3.13).

The scientific results obtained from this setup justified the effort, leading to the first evidence for the substructure of vibrational entrance-channel states. These results influenced a change of emphasis in fission research: from studying the properties of fission fragments (conducted at nuclear reactors), to exploring the properties of the highly deformed fissioning nucleus (conducted at accelerators). At the 1969 IAEA Conference in Vienna, numerous reports were presented on (d,pf) reactions involving various nuclei, all conducted at tandem accelerators but exclusively using Si detectors. There was consensus on the observation of "transmission resonances," i.e., vibrational states directly coupled to the fission degree of freedom during elongation. The significance of our measurements lay in the complete resolution of all local states coupled to these vibrational states and in determining their spins. This enabled us to conclude that the excitation energies were at most 2–3 MeV (as opposed to 5–6 MeV in the first minimum). However, this did not provide direct proof of the existence of the second minimum, nor did it indicate the extent of the deformation.

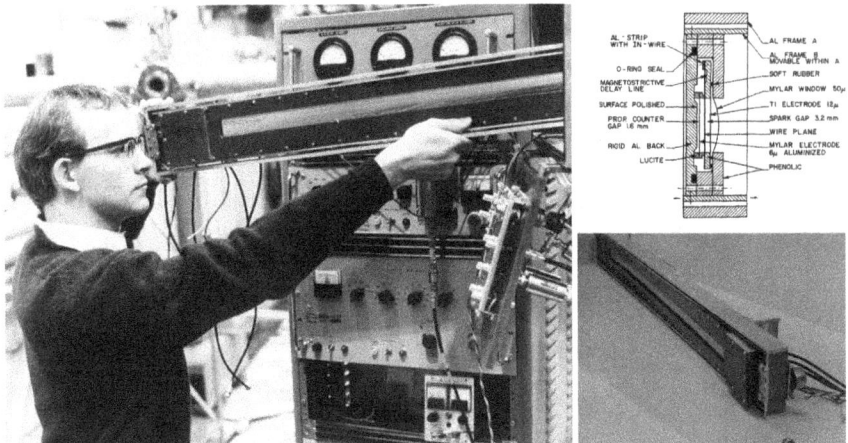

Fig. 3.11 Hans J. Specht assembling the novel spark chamber system © Hans J. Specht. All rights reserved. (upper right: transverse cross-section) © IAEA, 1969. All rights reserved. (Proceedings of the Second IAEA Symposium on Physics and Chemistry of Fission, Vienna, 28 July–1 August 1969)

Fig. 3.12 Spark chamber setup for magnetic spectroscopy—J.C.D. Milton, J.S. Fraser, and Hans J. Specht. Milton was head of Nuclear Physics, later the Physics Division, at the EN Tandem in Chalk River. © Hans J. Specht. All rights reserved

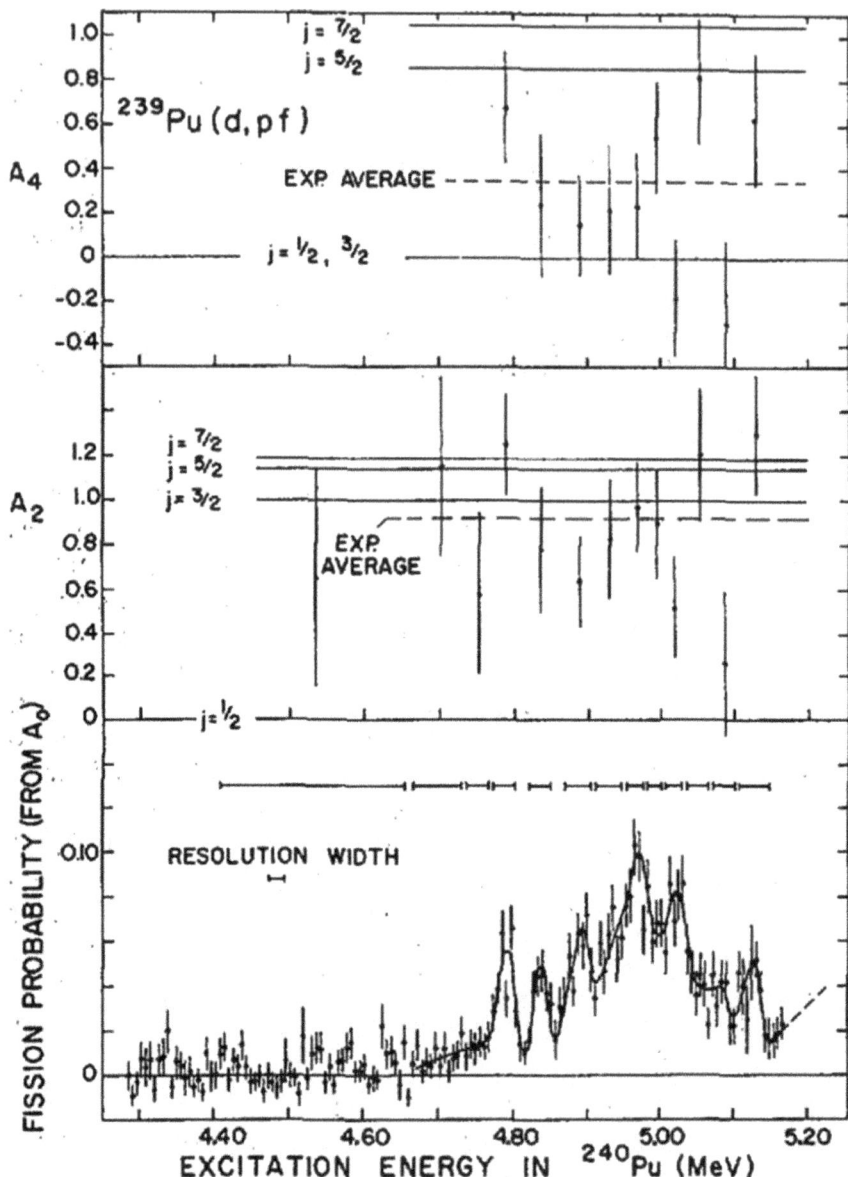

Fig. 3.13 Fission probability (from A_0) and angular correlation coefficients A_2 and A_4 were analysed to determine the spins of excited states through the angular correlation of fission fragments. For example, the first strong isolated line, at 4.792 MeV, suggests pure $j = 5/2$, leading to the assignment of spin 2. © IAEA 1969. All rights reserved. (Proceedings of the Second IAEA Symposium on Physics and Chemistry of Fission, Vienna, 28 July–1 August 1969)

Fig. 3.14 Left: MP Tandem accelerator facility in Garching; Right: Hans J. Specht and Ewald Konecny (from left to right). © Hans J. Specht. All rights reserved

Professor at Ludwig Maximilian University in Munich (LMU) 1969–1973

By early 1969, I was back in Munich. My colleagues in the Physics Department at Ludwig Maximilian University (LMU) considered the results from Canada and the associated detector developments, detailed in a comprehensive 100-page report (which was never published), to be sufficient for my habilitation in 1970. I could then quickly take on my own students and engage in several research topics in parallel at the new MP Tandem accelerator facility, part of the joint accelerator laboratory at (LMU) and Technische Universität München (TUM) on the new "campus" in Garching, near the FRM reactor. Some of this research was carried out in collaboration with E. Konecny from TUM. In 1971, I was appointed an H3 professor (Fig. 3.14).

A Key Experiment in 1972

Research topics included a significantly improved investigation of the ^{239}Pu(d,pf) reaction using the new Munich Q3D magnetic spectrograph, along with a specially constructed 2-m-long Charpak chamber (doctoral thesis by P. Glässel and diploma thesis by R. Männer, who later joined me in Heidelberg). Other research projects focused on measuring the triple-humped fission fragment mass distributions in the nuclide range of radium and actinium, investigating a possible octupole deformation at the second fission barrier (a doctoral thesis by J. Weber, which remains unique to this day). A particularly noteworthy highlight was the first measurement of a ground-state rotational band in the second minimum of a fission isomer [PLB 41 (1972) 43, doctoral thesis by D. Heunemann]. I will focus here on the latter accomplishment.

Probing Fission Isomers in the Second Minimum

The core idea was quite simple. In the nuclide range around $Z = 94$ and $N = 146$, discussed here, a nuclear reaction either results in the prompt fission of the excited compound nucleus or in the emission of electromagnetic radiation from the residual nucleus in the first minimum (Fig. 3.15).

However, with a probability of less than 10^{-4}, the second minimum can also be populated, and the radiation is emitted in a time-correlated manner before the delayed isomeric fission occurs. In deformed even–even nuclei, the de-excitation ultimately happens through E2 transitions to ground-state rotational bands with a spin sequence of 0^+, 2^+, 4^+, 6^+, and so forth. The moments of inertia describing these bands in both the first and second minima provide a direct signature of the different deformations. Since the transition energies are very low, all transitions are fully converted to conversion electrons. Therefore, conversion electrons must be measured instead of photons.

The idea came to me as early as 1970. After an unsuccessful preliminary attempt at the FRM with the 100 ns fission isomer ^{236}U, a breakthrough for such an experiment was achieved at the tandem accelerator in 1971. Here, the fission isomer ^{240}Pu was produced via the ^{238}U$(\alpha,2n)$ reaction, making it an ideal candidate due to its remarkably short half-life of just 4 ns. The target was placed at the center of a small ring-shaped silicon detector. Due to the recoil, the resulting isomers decayed in-flight by delayed fission within a few millimeters in front of the detector. This method ensured that only fission fragments from isomeric fission were detected, excluding the fragments from prompt fission, which were more than 10^4 times more frequent. The momentum spectrum of the electrons was measured in delayed coincidence using a non-magnetic β-spectrometer, which had been developed by E. Moll and E. Kankeleit during the early Maier-Leibnitz era.

For comparison, the relatively unknown rotational band in the first minimum was also measured. Sincere compliments are due here to the Munich accelerator team who, after an unforgettable lecture by me and only the first data had been taken, granted us a beam time quota several times higher than we were entitled to for 1971 in recognition of the importance of the measurement.

A Groundbreaking Discovery—Experimental Proof of Shape Isomerism

The proof of shape isomerism was considered one of the most important discoveries in the shell model since the introduction of spin–orbit coupling. The final result, shown in Fig. 3.16, revealed a significant difference in the rotational parameters of the two bands, providing the first experimental evidence for the existence of shape isomerism in atomic nuclei. Within the accuracy of the theoretical description of these parameters, the observed difference was consistent with a 2:1 deformation in the second minimum. This thereby proven generalization of the shell model, incorporating the symmetries of 2:1, 3:1, ... deformations, was considered one of the most important discoveries in the model since the introduction of spin–orbit

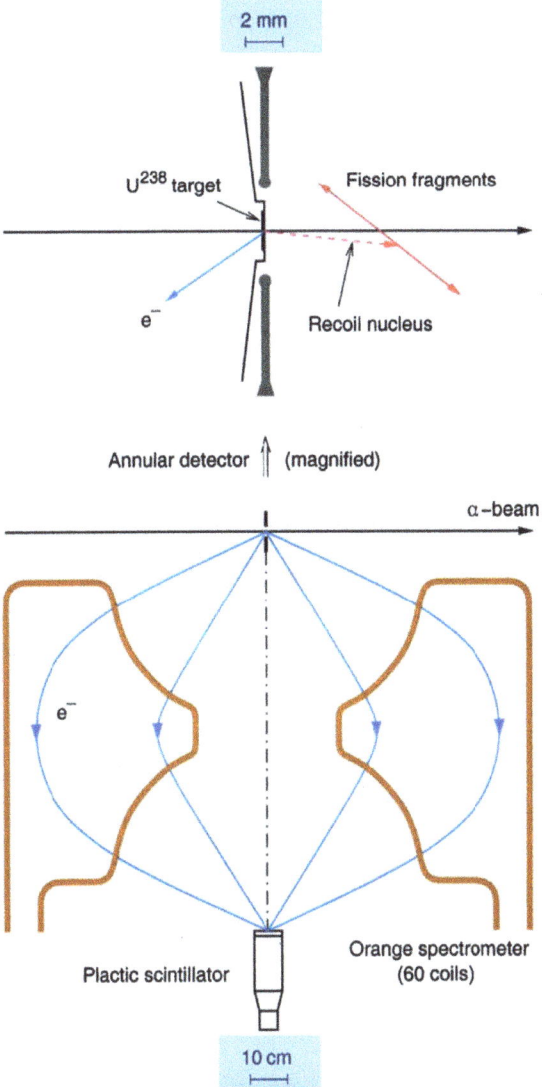

Fig. 3.15 Ring-shaped silicon detector with the target placed in the central aperture. Conversion electrons are measured in the orange β-spectrometer (lower part). Delayed coincidences are measured between a plastic scintillator in the focus position of the spectrometer and an annular semiconductor detector (upper part), detecting fragments from the decay of isomeric recoil nuclei in flight. © 1980, Elsevier B.V. All rights reserved

coupling. Consequently, many colleagues consider this first experimental proof as perhaps the most important work of my career (Figs. 3.17 and 3.18).

Public and Scientific Response to the Discovery

The public response to these results in 1972 was far beyond anything I had previously experienced. An experiment of this kind had been widely deemed unfeasible beforehand, so the immense curiosity it sparked was extraordinary. I was invited to

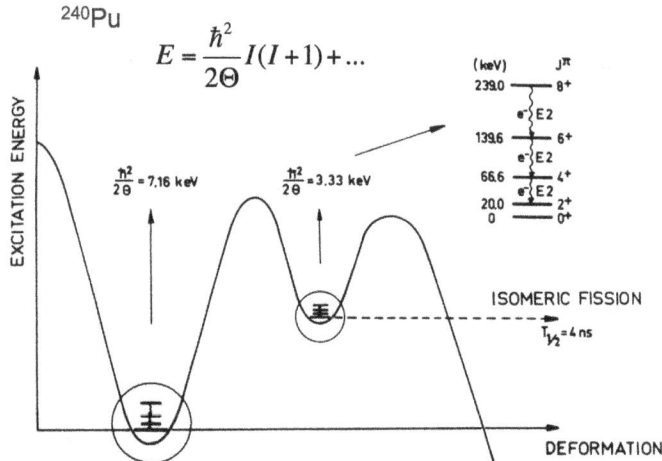

Fig. 3.16 Experimentally observed transitions within rotational bands in the first and second minimum (ground state and 4-ns fission isomeric state). The excitation energies and spins of the rotational levels are indicated. © Hans J. Specht. All rights reserved and © 1974 American Physical Society. All rights reserved

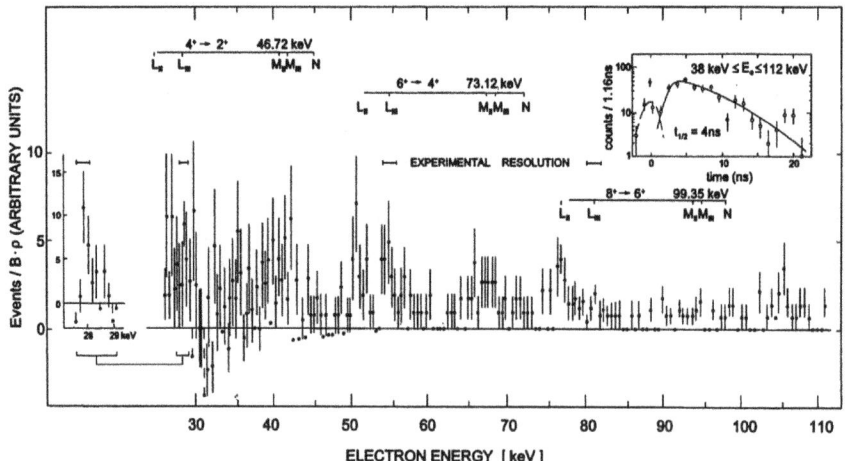

Fig. 3.17 Energy spectrum and decay curve (insert) of conversion electrons preceding isomeric fission of ^{240}Pu. © 1972 Elsevier B.V. and © 1980 Elsevier B.V. All rights reserved

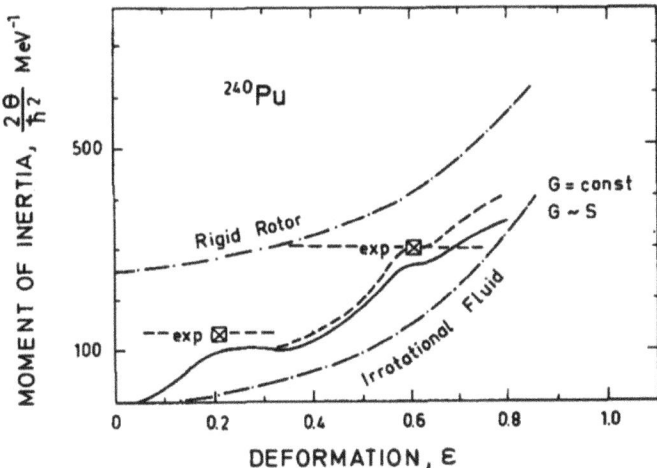

Fig. 3.18 Measured moments of inertia associated with rotational bands in ^{240}Pu compared to theory. © 1980 Elsevier B.V. All rights reserved

numerous lectures to present my findings, including one in Copenhagen, where I met A. Bohr and B. Mottelson (Nobel Prize laureates), and, one in Dubna, where I had the opportunity to meet G.N. Flerov, Y.T. Oganessian, and especially S. Polikanov, who was deeply affected by the findings. Over the years, there were many further meetings with all of them, though not all are remembered fondly due to Polikanov's later forced emigration (which ultimately found a positive resolution in the GSI/Heidelberg area). I still keep the thick folder of these invitations as a cherished reminder of this extraordinary time in my career. In 1973, I received many offers of professorships.

Notably, the invitation to a Heidelberg colloquium during the summer semester of 1972 was particularly significant and undoubtedly contributed to my prompt appointment as a full professor in Heidelberg—a decision I have never regretted.

Transition to Heidelberg in 1973—Full Professor

After accepting the offer for the winter semester of 1972/73, I got off to a flying start in Heidelberg. V. Metag at the MPI was enthusiastic about the prospect of collaborating with me, especially following his measurements on the lifetime systematics of fission isomers in Copenhagen. D. Habs, then close to completing his Ph.D. at the Haxel Institute, was so keen to use the new possibilities that he participated in several beam times in Munich even before my official move. P. Brix, newly appointed as director at MPIK, did everything possible to facilitate my group's access to the MP Tandem accelerator and to encourage collaboration with local colleagues. H.C. Pauli, who had also accepted an offer from MPIK following his theoretical work on the shell model, also joined the team (Fig. 3.19).

Fig. 3.19 Full professor at Heidelberg University. Experimental physics introductory lecture, main lecture hall, Philosophenweg 12, winter semester 1972/73. © Hans J. Specht. All rights reserved

From 1973 to 1983—Four Research Groups, Each Focusing on Different Areas

The following years fully met expectations. From 1973 to 1983, I had four research groups at the MPI Heidelberg and GSI Darmstadt, each focusing on different areas (Fig. 3.20):

- The first group, with D. Habs and V. Metag, undertook an extensive program at the MPIK Tandem, investigating fission isomers and transmission resonances below the fission barriers. Later, the group extended its research to include Coulomb fission experiments at the GSI UNILAC facility.
- The second group, with P. Glässel and D. von Harrach, initially focused on developing and constructing large position-sensitive parallel plate detectors for a future large-scale experiment at UNILAC. Their research centered on three- and four-body decays in nuclear collisions (see "Heavy-Ion Reactions").
- The third group, with R. Männer et al., specialized in hardware informatics, building, among other projects, the multi-processor system Polyp for UNILAC and a systolic array with 28 000 processors to function as a trigger processor for our later CERES experiment at CERN.
- The fourth group, with R. Schuch et al., focused on inner-shell ionization in atomic collisions, as already discussed in the research chapter on atomic physics.

Fig. 3.20 Research was conducted at the Tandem accelerator MP-5 at the Max Planck Institute (first beam in 1967), the UNILAC accelerator at GSI (first beam in 1974), and Heidelberg University's Physikalisches Institut. During this decade (1973–1983), approximately 30 diploma and Ph.D. students contributed to the research effort. © Hans J. Specht. All rights reserved

All six group leaders completed their habilitations at our faculty between 1974 and 1986 and subsequently went on to do further research (Fig. 3.21). In this decade, I supervised approximately 30 diploma and Ph.D. students.

Over the years, we also hosted many visitors, including several Humboldt scholars, each of whom stayed for a full year and occasionally returned for additional visits. Among them were C.O. Wene from Lund University, J. Wilhelmy from Los Alamos, P. Paul from Stony Brook University, J. Pedersen from NBI Copenhagen, S. Kapoor from BARC Bombay, and, finally L. Grodzins from MIT Boston.

I was particularly involved in the work of the first two groups, though less so in atomic physics due to the significant outside contributions. The working atmosphere was very personal and constructive, and I look back on these years with satisfaction and gratitude. It was a period marked by mutual exchange and collaboration, especially with the many diploma and doctoral students involved at that time.

The work of the Habs/Metag group on the spectroscopic properties of fission isomers received special attention, going significantly beyond the Munich results. Their research included combining spin and fission isomerism within the same nucleus, studying spins and magnetic moments in odd nuclei, examining rotational bands with measurements of moments of inertia in nuclei beyond ^{240}Pu, and even measuring lifetimes of rotational states to determine quadrupole moments. The latter used the "charge plunger technique" (CPT), based on an idea credited to D. Habs (Fig. 3.22). This technique involved measuring highly-converted transitions within rotational bands, which resulted in cascades of Auger electrons due to the filling of

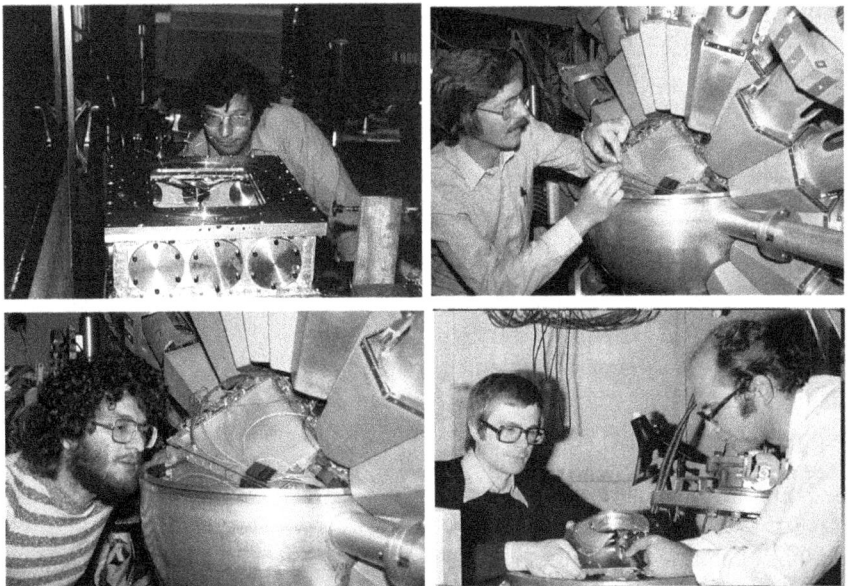

Fig. 3.21 D. von Harrach, P. Glässel, J. Schukraft, V. Metag, and D. Habs in the 1970s (top left to bottom right). © Hans J. Specht. All rights reserved

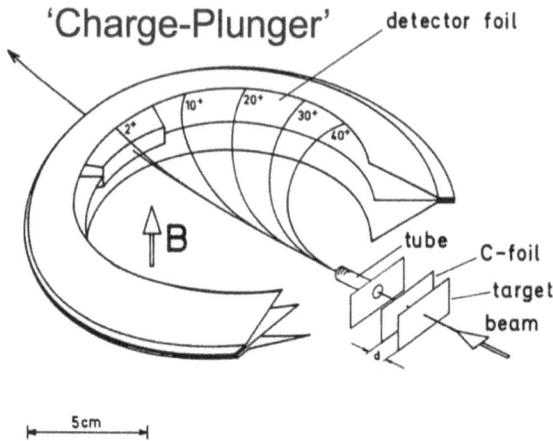

Fig. 3.22 Experimental setup known as the charge plunger for measuring charge distributions as a function of the distance between target and carbon foil. © 1977 American Physical Society. All rights reserved

multiple vacancies in the inner electron shells, leading to very high charge states exceeding 40+.

These high charge states can be reset to equilibrium through electron capture by placing a thin foil in the path of the recoiling fission isomers, provided the transitions occur before the foil is reached. By measuring the fraction as a function of the target–foil distance, the mean lifetime of the rotational transitions could be determined within a time range of 0.1–1 ns.

The charge distribution was measured by deflecting recoil nuclei in a magnetic field towards a macrofoil detector, where the atoms were stopped and the resulting delayed fission traces were recorded. The quadrupole moments of the fission isomers (Fig. 3.23) were approximately three times larger than those of the ground states. The nearly model-independent correlations with the axis ratio of the fission isomers yielded an average value of 2.0 ± 0.1 for the isomers, compared to approximately 1.3 for the ground states.

Another remarkable result from the group was the discovery of Coulomb fission. This process was analogous to Coulomb excitation, which has been used for many years in nuclear structure physics, but it required heavy projectiles with high atomic numbers due to the approximately 6 MeV fission barriers in actinides. Such projectiles became available at the UNILAC facility at GSI Darmstadt from 1976. The process was particularly intriguing from a theoretical standpoint because the Coulomb interaction directly excited the "doorway" states leading to fission, a concept extensively advocated by the Frankfurt School of W. Greiner et al.

In an initial attempt using silicon (Si) detectors (as part of J. Schukraft's diploma thesis), a nuclear contribution, such as nucleon transfer, could not be ruled out. This issue was resolved using the reaction ^{238}U+^{184}W, with clean detection of the intact recoiled ^{184}W projectiles by observing their lowest excited 2^+ state via decay (conversion) electrons. Both the shape and magnitude of the resulting excitation function aligned with theoretical expectations over many orders of magnitude.

Fig. 3.23 Quadrupole moments of fission isomers and ground states of actinide nuclei in comparison to theory [Phys. Reports 65 (1980) 1]. © 1980 Elsevier B.V. All rights reserved

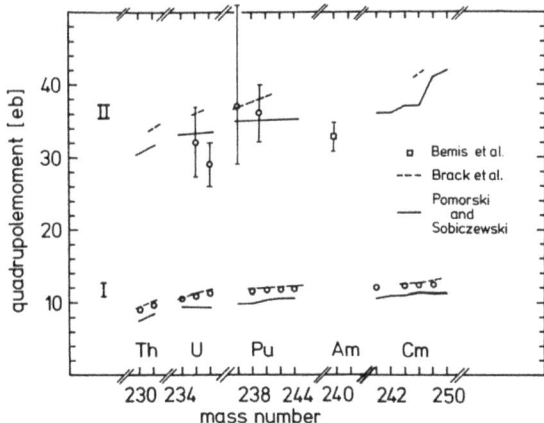

Selected Publications—A Guide

Nuclear Physics: Key Open Questions in Nuclear Fission

- Prompt and Delayed Gamma Rays from Fission

  H. Maier-Leibnitz, P. Armbruster, and H.J. Specht

  Physics and Chemistry of Fission, IAEA, Vienna (1965) II, 113–123

  Oral presentation by H. Maier-Leibnitz. My hardware contributions and the general spirit in P. Armbruster's group led to co-authorship of several papers on nuclear fission and paved the way for future developments.

Fission Isomers and the Double-Humped Fission Barrier

- A High Resolution Study of the ^{239}Pu (d,pf) Reaction

  H.J. Specht, J.S. Fraser, J.C.D. Milton, and W.G. Davies

  Physics and Chemistry of Fission, IAEA, Vienna (1969) 363–373

  First transmission-resonance spectroscopy of the double-humped fission barrier, made possible by replacing the then-common photo emulsions in magnetic spectrographs by a spark chamber (part of my habilitation thesis).

- Intermediate Structure in the ^{239}Pu (d,pf) Reaction

  P. Glässel, H. Rösler, and H.J. Specht

  Nucl. Phys. A256 (1976) 220–242

  Greatly improved with the Munich Q3D spectrograph and a multiwire chamber, since invented by G. Charpak (Ph.D. thesis P. Glässel).

- Identification of a Rotational Band in the ^{240}Pu Fission Isomer

  H.J. Specht, J. Weber, E. Konecny, and D. Heunemann

  Phys. Lett. 41B (1972) 43–46

  Pioneering paper giving experimental proof of nuclear shape isomerism with a 2:1 deformation (detection of fission isomers by S. Polikanov et al. in 1962 and generalized nuclear shell model by V. Strutinsky).

- The Quadrupole Moment of the 8 μs Fission Isomer in ^{239}Pu

  D. Habs, V. Metag, H.J. Specht and G. Ulfert

  Phys. Rev. Lett. 38 (1977) 387–389

  First quadrupole moment of a fission isomer (charge plunger technique).

- Spectroscopic Properties of Fission Isomers

  V. Metag, D. Habs, and H.J. Specht

  Physics Reports 65 (1980) 1–41

  Review of the unique set of results on fission isomers in Heidelberg.

Fission Fragment Mass Distribution

- Symmetric and Asymmetric Fission of Ac Isotopes near the Fission Threshold

  E. Konecny, H.J. Specht, and J. Weber

  Phys. Lett. 45B (1973) 329–331

  First of a series measuring triple-humped fragment mass distributions and their relation to octupole deformations in the late fission process.

- Nuclear Fission

  H.J. Specht

  Rev. Mod. Phys. 46 (1974) 773–787

  A landmark publication in the field and a vital reference for understanding the complexities of nuclear fission research.

Heavy-Ion Reactions

Entering the Realm of Heavy-Ion Physics

The commissioning of the UNILAC heavy-ion accelerator at the Gesellschaft für Schwerionenforschung (GSI) in Darmstadt in the mid-1970s led me into a third area of research: the physics of heavy-ion interactions at energies above the Coulomb barrier. The first experiment with U+U collisions in 1976 was conducted by a small GSI group led by R. Bock, using a straightforward setup with Si detectors. Both D. von Harrach and I were deeply involved in the project. However, our hopes of advancing into the realm of superheavy elements through nucleon transfer were soon to be dashed. Instead of observing a symmetric distribution of resulting nuclei around U ($Z = 92$), we observed a broad continuum with a yield reduced by a factor of 10^3 for nuclei with $Z < 70$. Beyond U, the spectrum was cut off almost vertically at $Z = 96$, the yield falling off by more than a factor of 10^5, clearly reflecting the high fission probability in the actinide region.

The Harrach/Glässel group was at the center of a major experiment established at UNILAC, in collaboration with MPIK (H. Sann). The experiment featured an impressive setup using two large, 1 m^2 position-sensitive parallel-plate gas detectors for time-of-flight and dE/dx measurements, paired with an equally large ionization chamber. All components were freely movable within an evacuated chamber of diameter 3 m and height 4 m, known as the "Heidelberger Fass." This pioneering configuration enabled groundbreaking measurements of three- or four-body decays in collisions of heavy nuclei up to U+U, at a time when most other experiments were still relying on semiconductor detectors (Fig. 3.24).

Fig. 3.24 Experimental setup at GSI known as the 'Heidelberger Fass'. © Hans J. Specht. All rights reserved and © 1979 Elsevier B.V. All rights reserved

Fig. 3.25 Hans J. Specht, always a sharp voice of inquiry, engaging the audience with his characteristically thought-provoking questions at an early conference. © Hans J. Specht. All rights reserved

Heavy-Ion Induced Fission and "Break-up" Processes

The main topic of these investigations had a well-established background, centered on "deep-inelastic" collisions of light to medium-heavy nuclei. These collisions demonstrated an astonishingly large transfer of energy and angular momentum between the collision partners, described theoretically using statistical models ("diffusion"). The observation of more than two-body decays in the final state provided insights into the magnitude of the angular momentum transfer through angular correlations. Additionally, we gained a deeper understanding of the collision dynamics, such as local timescales ("proximity effects"), which differentiate between breakup and decay in statistical equilibrium, such as in sequential fission (as reviewed in *Phys. Bl. 37, 7 (1981) 199*). At the time, this was groundbreaking, filling the agendas of specialized conferences and resulting in a series of Physics Review Letters. However, despite the importance and systematic nature of these results, they seemed to resonate mostly within a specialized scientific community. Even the influx of new diploma students—always a sensitive indicator of how compelling a field is—no longer met expectations. For me, this signaled that a change was in the air, much to the regret of many colleagues involved in this research area (Fig. 3.25).

Selected Publications—A Guide

Heavy-Ion induced Fission and "Break-up" processes

- Fission of ^{238}U Induced by ^{136}Xe for Energies Close to the Coulomb Barrier

  D. Habs, V. Metag, J. Schukraft, H.J. Specht, C.O. Wene, K.D. Hildenbrand

  Z.Phys. A283 (1977) 261–268

- Direct Observation of Coulomb Fission of ^{238}U with ^{184}W Projectiles

  H. Backe, F. Weik, P.A. Butler, V. Metag, J.B. Wilhelmy, D. Habs, G. Himmele, and H.J. Specht

  Phys. Rev. Lett. 43 (1979) 1077–1080

  Detection of Coulomb fission, analogous to Coulomb excitation in nuclear structure. First evidence in the 1977 paper (diploma thesis J. Schukraft).

- A Square Meter Position Sensitive Parallel Detector for Heavy Ions

  D. von Harrach and H.J. Specht

  Nucl. Instr. Meth. 164 (1979) 477–490

  Part of a special installation at GSI's UNILAC in Darmstadt, called the "Heidelberger Fass" in the local jargon. An evacuated container, several meters in diameter and height, housed freely movable large-area parallel-plate detectors and a large ionization chamber (built by H. Sann). The setup allowed for kinematically complete (exclusive) measurement of 3 or 4 nuclei in the final state, at a time when small Si detectors were the common tool.

- Angular Momentum Transfer in Deeply Inelastic Collisions from Exclusive Sequential-Fission Experiments

  D. von Harrach, P. Glässel, Y. Civelekoglu, R. Männer, and H.J. Specht

  Phys. Rev. Lett. 42 (1979) 1728–1732

- Three-Particle Exclusive Measurements of the Reactions ^{238}U+^{238}U and ^{238}U+^{248}Cm

  P. Glässel, D. von Harrach, Y. Civelekoglu, R. Männer, H.J. Specht, J.B. Wilhelmy, H. Freiesleben, and K.D. Hildenbrand

  Phys. Rev. Lett. 43 (1979) 1483–1486

- Direct Observation of Proximity Effects in Ternary Heavy-Ion Reactions

  D. von Harrach, P. Glässel, L. Grodzins, S.S. Kapoor, and H.J. Specht

  Phys. Rev. Lett. 48 (1982) 1093–1097

- Direct Observation of Non-Equilibrium Effects in Sequential Fission

  P. Glässel, D. von Harrach, L. Grodzins, and H.J. Specht

  Phys. Rev. Lett. 48 (1982) 1089–1093

  Several independent topics from well-defined exclusive final states.

- Summary of the Symposium

  H.J. Specht

  Physics and Chemistry of Fission 1979, Juelich, Proceedings IAEA Wien (1980) II, 459–476

  A formal end to an era, with papers for all the symposia (1965, 1969, 1973, 1979).

Research at the CERN Accelerators from 1983 onwards

Ultra-Relativistic Heavy-Ion Physics— Searching for the Quark–Gluon Plasma

My "Phase Transition" to the Phase Transition

A transformative change came about almost by chance—another peculiar coincidence in my life. At that time, discussions emerged about launching and exploring the exciting new field of "quark matter in the laboratory" through ultra-relativistic heavy-ion collisions at one of CERN's accelerators. I was drawn into these discussions around 1979 and became deeply involved in these pioneering efforts. From 1983 until about 2010, I played an active role in four major experiments at CERN, three of which were exclusively dedicated to this new and groundbreaking field:

- **Axial Field Spectrometer, R807/R808** at the ISR,
- **HELIOS NA34-2**,
- **CERES/NA45**, and
- **NA60** at the SPS.

Pioneering Role in High-Energy Heavy-Ion Experiments

In two of these experiments, I served as spokesperson—first for the 1st generation heavy-ion experiment HELIOS NA34-2, and later as the founder and spokesperson of the 2nd generation heavy-ion experiment CERES/NA45. I also contributed to the early conceptual design of an LHC heavy-ion experiment, providing foundational input for the initial phase of the ALICE experiment and continuing to contribute in an advisory capacity during its early development (Fig. 3.26).

The Birth of Heavy-Ion Physics at CERN's SPS

The decade-long discussion process was a prime example of the constructive interplay between the low-energy and high-energy physics communities. It showcased how the interests of various laboratories and individual accelerator plans—at

Fig. 3.26 A new era from 1983. Heavy-ion physics at CERN: forty years of commuting between Heidelberg and Geneva. © CERN. All rights reserved

Lawrence Berkeley Laboratory (LBL), GSI, and CERN—were harmoniously integrated, thanks to the dedication of individuals who made this their personal mission. The matrix shown in Fig. 3.27 attempts to organize the historical flow of important milestones and foundational steps which ultimately led to the beginning of the heavy-ion program at CERN. Milestones highlighted in red indicate points where my contributions were particularly significant. This matrix originally came from a colloquium I gave at CERN in 2016 as part of CERN's 60th anniversary celebration, which covered the field of heavy-ion physics ("30 Years of Heavy Ions: … What next?").

Collaborative Beginnings: GSI Darmstadt and LBL Berkeley Join Forces (1974)

Just in time for the commissioning of the BEVALAC in 1974, a collaborative agreement was established between GSI Darmstadt and LBL Berkeley, with R. Bock and H. Grunder as the driving forces. The goal of exploring nuclear collisions at 1 GeV/u was to study compressed nuclear matter, aiming to gain insights into the equation of state and its relevance to the interiors of neutron stars. Initially, this research was grounded solely in hadronic theories. A year later, in 1975, the idea of the possible existence of a really new state of matter spread within the community. The first publication based on partonic physics appeared, suggesting the possibility of "deconfinement" of quarks and gluons—which are normally confined within hadrons—under conditions of both high temperatures and high (net) baryon densities.

The significantly higher energies required for these studies, beyond what was available at LBL, were accessible for proton–proton (pp) collisions at Fermilab, as well as at the SPS and, especially, at the ISR at CERN. These opportunities were swiftly exploited, sparking increased interest in the topic—though initially much more so among particle physicists than nuclear physicists, due to the need for much higher energies than were available at the Bevalac. In 1977, Fermilab's data on muon pair production in the invariant mass range of $1 < M < 3$ GeV were interpreted by

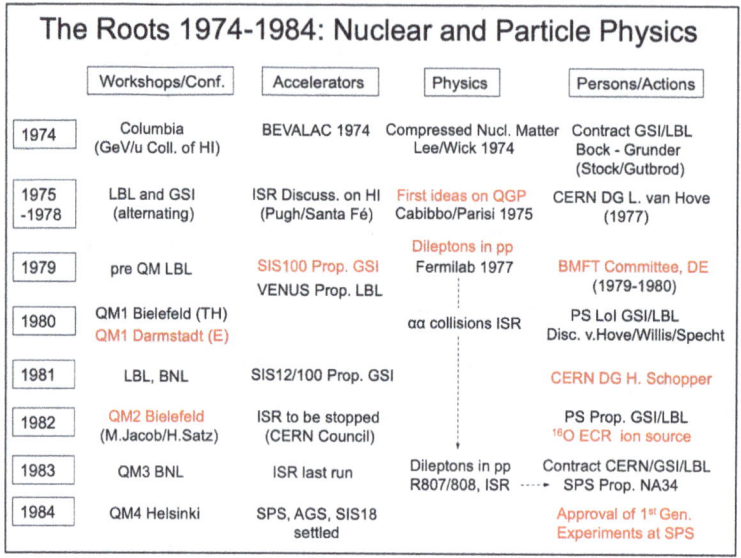

Fig. 3.27 The birth of heavy-ion physics at CERN. © Hans J. Specht. All rights reserved

E. Shuryak in 1978 as thermal radiation from a state formed in pp collisions. Though the data were problematic, they served as milestones in theoretical interpretations, introducing the term "quark–gluon plasma" (QGP), which is now commonly used and was first coined in the literature. At the ISR, a beam time with αα (alpha–alpha) collisions took place in 1980. However, only hadronic observables were measured, and no significant new findings emerged. Although the idea of accelerating heavy ions at the ISR was proposed as early as 1975 and dominated discussions at CERN for the following 5–6 years, the historical trajectory ultimately took a different direction.

The Road to CERN: Discussions, Decisions, and Breakthroughs

In 1979, GSI in Darmstadt proposed its own project for the heavy-ion synchrotron SIS100, aiming at the energy range of CERN's PS, with ions up to 10 GeV/u for uranium. However, the dilemma with this proposal was that the energy was too low for the clear formation of a quark-gluon plasma (QGP), making it essentially a continuation of hadronic research, though employing more radical approaches. Furthermore, the proposal did not reflect the interests of the broader GSI community, which had significantly different priorities, leading to substantial internal conflicts.

Ad-hoc Committee on Nuclear Physics of the German BMFT

In response, the Federal Ministry for Research and Technology (BMFT), under the leadership of H. Lindenberger, established an "Ad-hoc Committee on Nuclear Physics" in late 1979 to evaluate all project proposals from GSI, Jülich, Munich, and other institutions, thereby shaping the future direction of the field in Germany. As a member of this committee, and having developed a keen interest in experiments at sufficiently high energies to address deconfinement, I saw no alternative to opening this new area at CERN. I witnessed the unanimous recommendation reached by all six committee members in June 1980: "To explore the possibility of developing ultra-relativistic heavy-ion research at a CERN accelerator through a collaboration between CERN and GSI."

A Milestone—The First Organized Discussion between Particle and Nuclear Physicists

This led to a chain reaction for me personally. Just a few months later, in October 1980, R. Bock and R. Stock organized the "Workshop on Future Relativistic Heavy Ion Experiments" at GSI, now known as the first Quark Matter Conference, "Quark Matter 1," where, for the first time, experimental physicists from CERN (about 30% of the participants), the BEVALAC, and those involved in low-energy physics came together to engage in discussions with prominent theorists like Hagedorn. It was here that I first met William J. Willis, or as we all affectionately called him, Bill Willis, who made a lasting impression on me. The discussion was dominated by the dream of "keeping the ISR," as CERN was focused on the construction of the Large Electron–Positron Collider (LEP) and planning to shut down the ISR. I was assigned to deliver the "Summary Talk" at the end of the event, a considerable challenge given my background in low-energy nuclear physics. However, my enthusiasm for this rapidly evolving area of research surpassed any gaps in my expertise. Afterward, P. Brix, who was in the audience, congratulated me and urged me to write up the summary within the next 24 h—a task I successfully accomplished.

The opening lines of my summary read: "Standing up here I wonder how I ever got there. In case somebody has not realized that, I am strictly a low-energy nuclear physicist whose horizon, by my own experience, has never yet exceeded 12 MeV/u. Speaking about something one does not understand is already painful, but summarizing something one does not understand is truly disastrous. You must ask R. Stock what trick he played to get me on this panel. It was only today that I found out the real reason, and now I am grateful to him both for the trick, and of course for organizing this very successful and memorable meeting (Fig. 3.28)."

A Game-Changing Conversation

Shortly thereafter, Bill Willis arranged a lengthy discussion with the then Director-General of CERN, L. van Hove. This was a three-way conversation among Bill Willis, L. van Hove, and myself, an experience I still vividly recall. During this 2-h discussion with L. van Hove, I gained profound new insights, and my initial stance of "keep the ISR," which had been my guiding principle just a few weeks earlier, was replaced by a newfound understanding. I realized that, as a "fixed-target" machine, the CERN SPS offered luminosities a 1000 times greater than those achievable with the ISR. From that point on, resistance to the planned early shutdown of ISR in favor

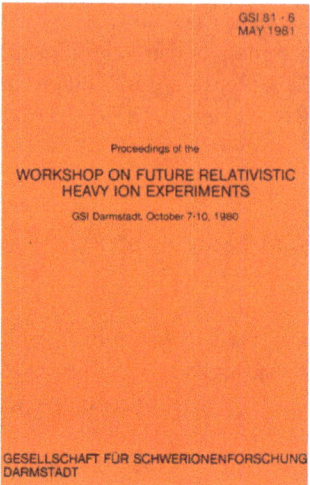 

Fig. 3.28 First Quark Matter Conference, 1980 at GSI. Editors of the Proceedings: R. Bock and R. Stock. Summary speaker: Hans J. Specht. *Hans J. Specht's final slide and concluding statement in his write-up carried the bold declaration: 'The Future has already begun.'* © GSI (left panel), © FAIR-GSI, with a photo credit to G. Otto (right panel). All rights reserved

of constructing LEP came only from the ISR community, which had just begun new pp experiments. L. van Hove continued to express his support for the SPS. I left CERN with a clear decision to immerse myself in this new field.

In 1981, H. Schopper became the Director-General of CERN. Shortly before this, G. zu Putlitz, the Scientific Managing Director of GSI, had given the green light for a proposal by the GSI/LBL group for two experiments at the CERN PS, covering the energy range of the SIS100. Initially submitted as a Letter of Intent (LoI) in early 1982, this proposal developed into a full-fledged plan, which also included an offer for a modern ECR + RFQ ion source to produce ^{16}O/^{32}S ions.

In mid-1982, the "Quark Matter 2" conference took place in Bielefeld, marking the first systematic discussions between particle and nuclear physicists on the theoretical and experimental aspects of quark-gluon plasma (QGP) formation in ultra-relativistic nucleus–nucleus collisions. The organizing committee, composed of key figures like T. Ericson, M. Jacob, H. Satz, and B. Willis, was now dominated by high-energy physicists who had taken on the role of pioneers in the field since 1980. Their influence was prominent, representing about 80% of the participants. Six working groups were established, each focused on defining fundamental principles for the first-generation heavy-ion experiments at the SPS. These principles were based on proposed observables up to that point, including lepton pairs across all mass regions. I served as a convener of one of these groups alongside S. Nagamiya (Working group #5: Inclusive Measurements and Particle Identification). The atmosphere among us was one of enthusiasm and optimism. S. Nagamiya later served as the keynote speaker at a GSI event celebrating my 60th birthday in 1996.

The GSI proposal for the PS initially played a somewhat peripheral role in this context. Negotiations with CERN were led primarily by R. Bock. When I joined the CERN PSCC in 1983, I became involved in the process ex officio. Approval for the PS was granted that same year, but it was subsequently put on hold to allow for further development of experiments originally designed for the SPS. By 1984, CERN had formally approved the proposed SPS experiments, including the adapted GSI proposals for the SPS.

The decision to launch an experimental heavy-ion program at CERN-SPS came shortly after the SPS enabled the discovery of the W and Z bosons in 1983, marking a new frontier for the facility. By 1987, a total of six heavy-ion experiments had been approved.

In October 1986, oxygen-16 ions were successfully accelerated by the SPS and fired into a fixed gold (Au) target, marking the beginning of the heavy-ion program. The experiments were designed with different detector systems, and each adapted to investigate different aspects of the quark-gluon plasma (QGP) phase transition. The six first-generation experiments included HELIOS NA34/2, the NA35 streamer chamber, NA36, the NA38 dimuon spectrometer, WA80 ("Plastic Ball"), and WA85.

The results from these initial experiments soon made it clear that the field required truly heavy ions, such as Pb, beyond ^{16}O and ^{32}S. In response, the concept for this was developed by N. Angert, H. Haseroth, and others by 1990 and implemented as "LINAC3." Commissioned at CERN in 1994, LINAC3 was part of an international collaboration led by GSI, along with CERN, GANIL, INFN Legnaro/Torino, and IAP Frankfurt. LINAC3 continues to operate at CERN to this day.

Regarding the role of the Director-General H. Schopper since 1981, it is thanks to his decision that heavy-ion physics became a permanent program at CERN, extending from the SPS to the present day at the LHC. With his wisdom and his ability to recognize quality beyond the mainstream of high-energy physics, he personally engaged in numerous discussions, both with specialized committees and with individuals.

The key contribution of the GSI/LBL group was not so much the physics or the expertise they brought, but rather their offer of the original ion source, which, along with the beam transport, enabled CERN to begin accelerating heavy ions almost cost-free. H. Schopper then tied this into a zero new contribution approach for experiments: there would be no additional funding from CERN, but experiments that were being wound down or major components of existing detectors could be repurposed for heavy-ion physics. He also personally made the final positive decision at the highest level. In his own words, it was "one of the three decisions in my life which I made against the advice of my own committees."

Note: There is a notable comment in the CERN Courier (April 2017), published after the special event "30 Years of Heavy Ions: … What Next?" as part of CERN's 60th anniversary celebration: *"As the then CERN Director-General Herwig Schopper recalled at the November workshop, the 1980s were not the best time to initiate new projects. The CERN budget was severely cut and the laboratory was very much focused on the construction of the Large Electron–Positron Collider (LEP). Despite this, Schopper bravely decided to give heavy-ion physics a chance. He was motivated by arguments put forth by Reinhardt Stock, Hans Specht, Rudolf Bock, Bill Willis, and several other leading physicists."*

Theoretical Background—Motivation for Ultra-Relativistic Heavy-Ion Collisions

Quantum chromodynamics (QCD), a fundamental component of the Standard Model of particle physics, underpins much of our understanding of the strong interaction—the fundamental force responsible for binding quarks and gluons (the elementary building blocks of matter) into hadrons, such as protons and neutrons, which together form atomic nuclei. Research using ultra-relativistic heavy-ion collisions aims to explore QCD under extreme conditions of high temperature and/or matter density and in the low momentum non-perturbative regime, where the strong interaction is very strong indeed.

At the heart of this research is the concept shown in Fig. 3.29. The two primary phases of strongly interacting matter are:

– quasi-free quarks and gluons at extremely high temperatures or baryon densities, and
– distinct hadrons such as baryons (e.g., protons and neutrons) composed of three confined quarks at lower temperatures and densities.

A first-order QCD phase transition likely separates these two phases—the quark-gluon plasma (QGP) and the hadron gas—over a specific range of temperatures and densities.

Following the Big Bang, approximately 13.8 billion years ago, the universe underwent complex early phases of continuous cooling and expansion, eventually reaching the state known as a quark-gluon plasma (QGP). This extreme phase probably lasted until a few microseconds (around 10^{-5} s) after the Big Bang, when a QCD phase transition would have occurred, leading to the confinement of quarks and gluons into

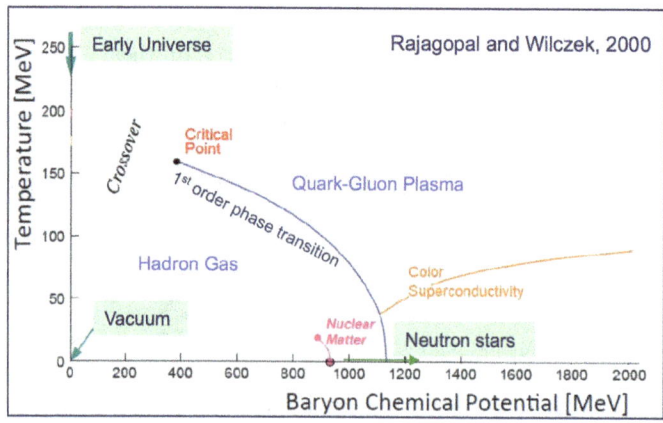

Fig. 3.29 The phase transition from a quark-gluon plasma (QGP) to a hadron gas. © K. Rajagopal and F. Wilczek (arXiv:hep-ph/0011333). All right reserved

hadrons. At this pivotal moment, empty space between these newly-formed structures was created for the first time, i.e., the vacuum was born.

This phase transition is accompanied by another fundamental property of QCD: chiral symmetry breaking. The Nobel Prize was awarded for this in 2008. Chiral symmetry breaking imparts mass to hadrons, making their mass approximately 50 times greater than that of the confined quarks, and is responsible for the origin of the masses of light hadrons. The protons, which emerged shortly after this phase transition, dominated the subsequent evolution of the universe and continue to define its observable mass to this day (Fig. 3.30).

Since the universe did not become transparent until approximately 370 000 years after the Big Bang (as evidenced by the cosmic microwave background), this early epoch remains beyond the reach of observational astronomy. However, quark matter can, in principle, be created experimentally in the laboratory through collisions of heavy ions at high energies, and in particular at energies far exceeding their rest mass. These collisions produce an extremely hot and dense "fireball" of fundamental particles, effectively recreating the extreme conditions in the universe during its first few microseconds, when it was a quark–gluon plasma (QGP).

The resulting fireball, which contains thousands of quarks and gluons in its hottest phase, evolves over time like a "mini Big Bang in the laboratory": it expands, cools, and eventually undergoes a phase transition (at a critical temperature) into hadronic matter. Once the QGP cools below the phase transition, the initial energy density of the hadronised medium remains high enough to allow for inelastic interactions, which modify the medium's "chemical" composition by changing the types of particles. These inelastic interactions cease at the chemical freeze-out temperature, at which point the particle composition becomes fixed. Elastic interactions can continue for a time, gradually diminishing until thermal freeze-out is reached, at which point the particle momenta are fixed.

During the expansion of the fireball, various particles are emitted, including hadrons such as baryons and mesons (quark–antiquark pairs), as well as photons and

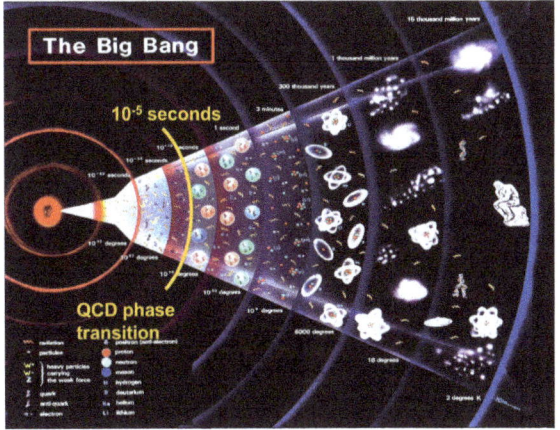

Fig. 3.30 Motivation for the heavy-ion research at ultra-relativistic energies. The early universe in the laboratory. © 1991–2025 CERN. All rights reserved

lepton pairs like electron–positron or muon–antimuon pairs. Each of these particles serves as an observable, providing insights into the state of the medium at different stages and capturing its complex evolution. A schematic illustration of "Quark matter formation at SPS energies: proof by circumstantial evidence" is shown in Fig. 3.31, offering a clearer understanding of the way the fireball evolves and highlighting the key probes for each stage in the process, including the intriguing quark–gluon plasma (QGP) and the chiral transition.

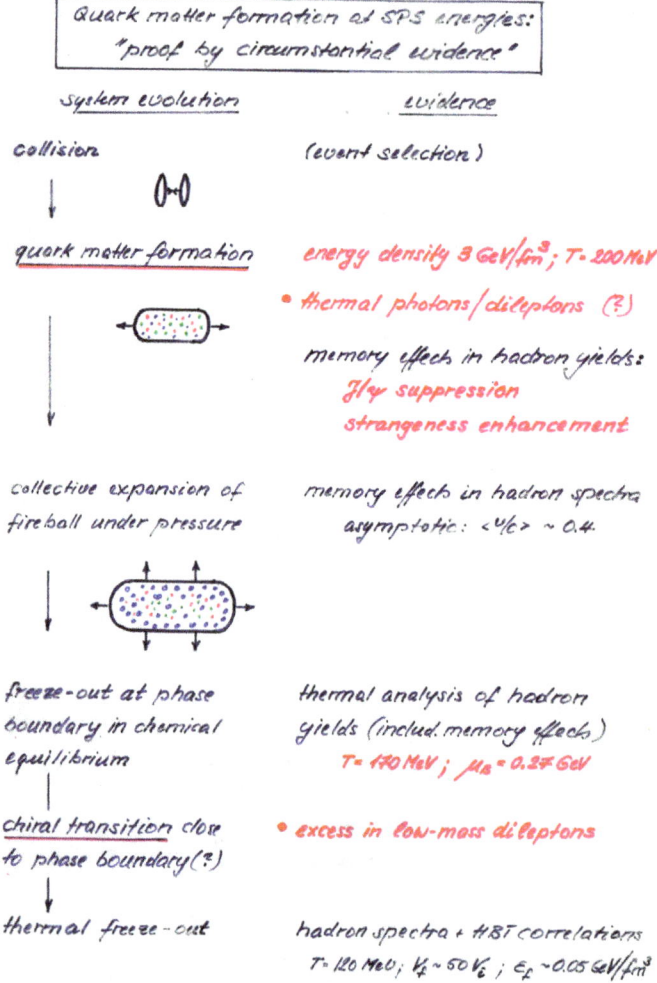

Fig. 3.31 A schematic illustration of quark matter formation at SPS energies, highlighting the key probes for each stage in the process. In Hans J. Specht's own writing. © Hans J. Specht. All rights reserved

In order to study these observables, one requires either large-scale universal experiments, such as those conducted today at the LHC, or a series of specialized experiments, as were carried out during the SPS era.

Legacy in Four Landmark Experiments at CERN—The Quark–Gluon Plasma Journey

Experiment R807/R808 at ISR

My first Scientific Associateship at CERN began in 1983 when I was invited by Bill Willis to join the R807/808 (Axial Field Spectrometer) experiment at the ISR, the world's first and, at that time, only hadron collider. As part of a team of 75 members, I had the unique opportunity to witness and contribute to the final year of operation of the ISR, immersing myself in high-energy physics as if I were a fresh student. My new environment, alongside prominent figures such as M. Albrow, I. Manelli, R. Palmer, B. Willis, and others, provided both intellectual stimulation and personal support, making integration a true joy. This first year at CERN, together with the international atmosphere, which was new to me, marked the second formative period of my life (after my time in Munich with H. Maier-Leibnitz), having a lasting and profound impact on me (Fig. 3.32).

The main components of the experiment were a central drift chamber and a (nearly) hermetic 4π uranium calorimeter. The primary focus was on jets and photons in the GeV range, while measurements of lepton pairs received special attention for future

Fig. 3.32 Axial Field Spectrometer (AFS) at ISR in October 1983. © 1983–2024 CERN, All rights reserved

planning. This interest led to a series of experiments in the decade following the discovery of the J/ψ particle by S. Ting in 1973 (Nobel Prize winner in 1976), sharpening the search for anything which might still have escaped detection. By 1983, there were already ten publications on individual leptons or lepton pairs beyond known sources, all within the domain of low transverse momenta and masses—coined "anomalous pairs."

First Measurement of e^+e^- Pairs with M < 1 GeV in pp Collisions at the ISR

The prospect of quark matter on the horizon further fueled my interest. As a newcomer, I was encouraged to explore this area and prepare an upgrade for the R808 setup to investigate these anomalies at the significantly higher pp collision energies of the ISR (63 GeV in the center-of-mass system). We decided to focus on electrons, which required us to install components for electron–hadron identification. My main responsibility was to build and integrate two large Cherenkov detectors, to be positioned between the NaI and U calorimeters and the drift chamber for electron identification.

The analysis of the data collected using this new setup was the subject of a doctoral thesis at Lund University, with J. Schukraft as the primary supervisor. After completing his own doctoral thesis in Heidelberg in 1984, Schukraft joined my GSI group and moved to CERN, where he played a pivotal role in this research. Our findings revealed a significant excess of lepton pairs at low masses, more clearly established than in previous studies. This result might be connected to the still unresolved puzzle of "soft photons" in pp collisions. J. Schukraft remained at CERN and later became the spokesperson for the ALICE collaboration for 20 years (Fig. 3.33).

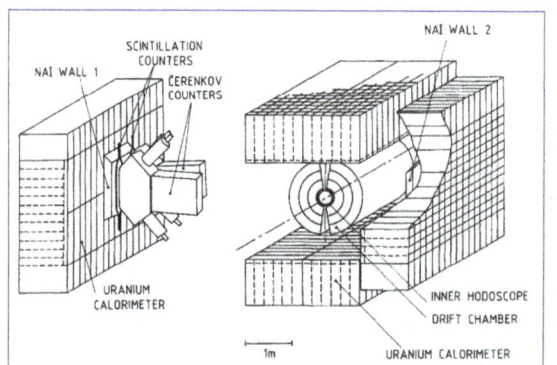

Fig. 3.33 My primary responsibility was the construction of Cherenkov detectors to upgrade the ISR R808 experiment, aimed at making the first measurements of lepton pairs in pp collisions. © 1987 Elsevier B.V. (left) and © Hans J. Specht (right). All rights reserved

Experiment NA34-2/HELIOS at SPS (Spokesperson)

While the ISR was still operational, Bill Willis led the conceptualization of a future experiment at the SPS: the High-Energy Lepton–Ion Spectrometer or HELIOS. This had two main objectives: HELIOS-1 aimed to investigate lepton production in proton–beryllium (pBe) collisions with a sensitivity 1000 times higher than that achieved at the ISR, while HELIOS-2 was designed to provide a comprehensive overview of the properties of nuclear collisions by incorporating additional components. N. McCubbin was appointed spokesperson for HELIOS-1, and I was selected as spokesperson for HELIOS-2, just in time for CERN's first round of proposals for heavy-ion experiments in 1984 (Fig. 3.34).

This role was far from what I had envisioned as a novice. Having a team of 150 members from 15 countries—where my ISR colleagues were in the minority—meant that managing human dynamics often played as significant a role as the physics we sought to explore. Shared components of the experiment included the U-calorimeter, adopted from the ISR experiment in a hermetic configuration, and the former NA3 muon-pair spectrometer.

HELIOS-1 featured a forward spectrometer equipped with a transition radiation detector (TRD), a uranium/liquid-argon calorimeter, and drift chambers for high-precision detection of electron pairs. HELIOS-2, on the other hand, used a magnetic spectrometer with particle identification capabilities for hadrons and, through external conversion, for photons as well. The acceptance of this spectrometer was determined by a slit in the U-calorimeter, enabling precise selection of the particles of interest.

HELIOS-1 achieved a resounding success within its designed scope. Its measurements of electron and muon pairs were significantly more precise than previous experiments, showing that the previously observed "anomalous pairs" (apart from those detected at the ISR) were likely artifacts arising from an incomplete understanding of the hadronic decays of mesons. Nonetheless, these earlier results marked important milestones in the theoretical interpretations.

HELIOS-2 successfully fulfilled its hadron-focused objectives, though it encountered challenges in the analysis of muon pairs. The calorimeter data, which covered

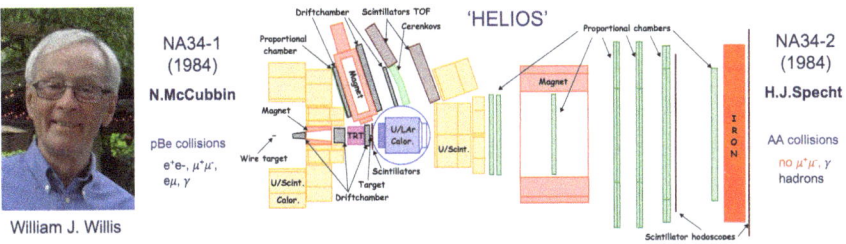

Fig. 3.34 Configuration of NA34/HELIOS-2 at CERN (with W.J. Willis, whom we all affectionately called Bill, leading the conceptualization). © CERN and © 1986 Elsevier B.V. All rights reserved

the full rapidity range from 0 to 6, provided exceptionally precise measurements of energy transfer in heavy-ion collisions. In sulfur–gold (S-Au) interactions, energy densities within the fireball reached 2.6 GeV/fm^3, well above the 1 GeV/fm^3 threshold considered necessary for quark-gluon plasma formation. This finding, combined with data from the NA35 experiment, was impactful enough to prompt a CERN press release in 1987, underscoring the contribution the experiment had made to understanding extreme conditions in nuclear matter (Figs. 3.35 and 3.36).

The analysis of hadron and photon data from the magnetic spectrometer was largely conducted by doctoral students from Heidelberg, but no direct evidence for the quark–gluon plasma (QGP) was found.

It soon became clear that the field was too complex to be fully addressed by a single experiment like HELIOS. The exciting prospect of lead (Pb) beams on the horizon accelerated the need for a more comprehensive approach. Within just 2 years, a unanimous decision was made to split the HELIOS collaboration and to pursue three separate experiments to explore the most intriguing observables in an optimal way: HELIOS-3, focusing exclusively on the muon pair spectrometer (led by G. London); NA44, optimized for hadron spectroscopy and HBT (led by H. Boggild); and NA45/CERES, dedicated to measuring electron pairs (led by me).

Fig. 3.35 First generation heavy-ion experiment at CERN: HELIOS NA34. © CERN. All rights reserved

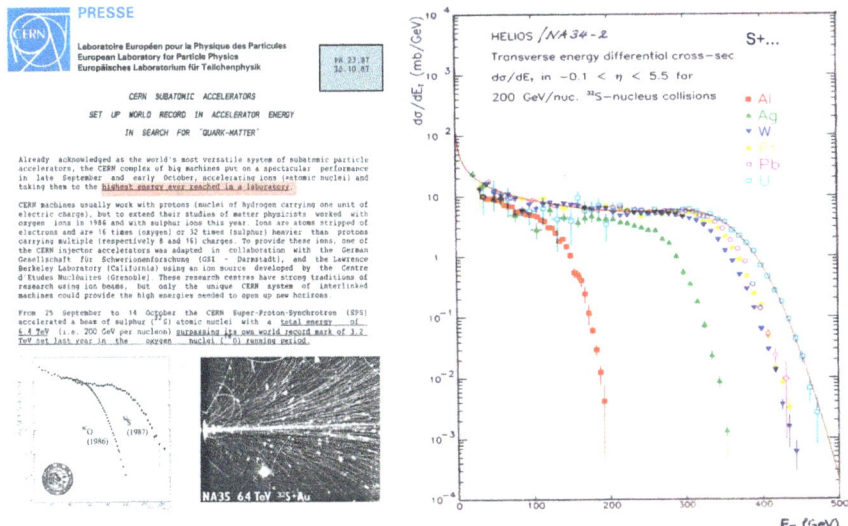

Fig. 3.36 CERN Press Release for Heavy-Ion Collisions: Results from HELIOS/ NA34-2 reveal transverse energy in 4π. The data for S–Au central collisions show an energy density of $\varepsilon = 2.6$ GeV/fm^3, surpassing the critical threshold of 1 GeV/fm^3 needed to form a quark-gluon plasma. © CERN. All rights reserved (left panel) and © 1988 Elsevier. All rights reserved (right panel)

Experiment NA45/CERES at SPS (Founder and Spokesperson)

The CERES "Spechtometer": A Masterpiece of Electron–Positron Spectrometry

While I had built many detectors in the past, this was the opportunity to design a full experiment. Technically, CERES was definitely the most adventurous experiment of my professional career. It was initially met with considerable skepticism from many colleagues who doubted it would ever work. Even the CERN Director-General at the time, Carlo Rubbia, personally came to judge my confidence before ultimately deciding to approve the project in 1989.

The experiment was based on ring-imaging Cherenkov (RICH) detectors for electron identification and tracking. Although the basic idea had already been included in the NA34-2 proposal, its implementation had initially been constrained by the limited space available in front of the calorimeter. The concept was later significantly optimized, largely thanks to support from Bill Willis.

The goal of the CERES/NA45 experiment was to measure electron pairs in a way which would remain compatible with future Pb beams. Lepton pair production is widely considered as a promising probe for studying the dynamical evolution of nuclear collision processes. The motivation for these measurements is as follows: in

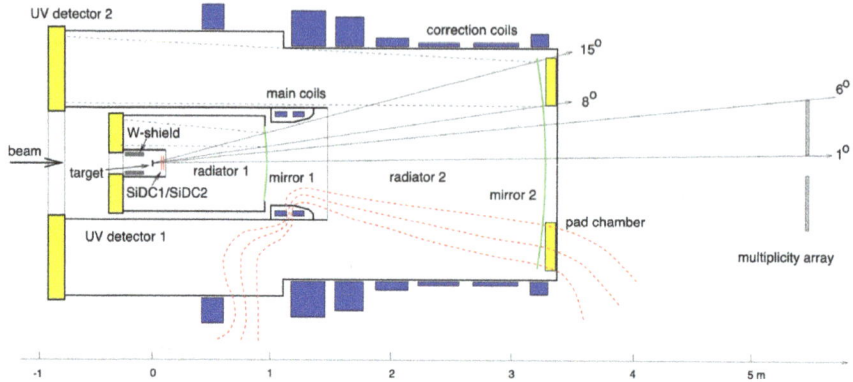

Fig. 3.37 Setup of the CERES/NA45 experiment at CERN SPS (used during the Pb-beam era). © 2005 Springer. All rights reserved

contrast to hadrons, leptons and photons are produced continuously throughout the entire space-time evolution of the fireball created in these collisions, starting from the early quark-gluon plasma (QGP) phase up to the final freeze-out of hadrons, when all interactions cease. Because leptons and photons are limited to electromagnetic interactions, they can escape the system without significant final-state interactions, as their mean free path is considerably greater than the size of the collision volume (Fig. 3.37).

CERES was the only experiment dedicated to measuring low-mass e^+e^- pairs emitted in nucleon- and nucleus-induced collisions at SPS energies. In the invariant mass range below 1 GeV—which was the only range accessible to CERES due to luminosity constraints—thermal dilepton production was dominated by the hadronic phase and mediated mainly by the light vector meson ρ (770 MeV). With its strong coupling to $\ell^+\ell^-$ and a lifetime of only 1.3 fm, much shorter than that of the resulting "fireball," the "in-medium" properties of this meson are highly sensitive to the chiral transition, making it a key test particle for studying this fundamental QCD phenomenon. These measurements pose significant challenges, including the low production probability of dileptons—about 10^{-4} compared to hadrons (i.e., 10 000 times smaller)—and a large background from unrecognized π-Dalitz and γ-conversion pairs (Fig. 3.38).

The basic principle of the CERES experimental setup was "hadron-blind tracking," achieved by using two ring-imaging Cherenkov (RICH) detectors with gas radiators: one positioned before and the other after a short superconducting double solenoid. The Cherenkov thresholds were set high enough to ensure that only electrons could produce Cherenkov rings. The momentum of the electrons was determined by the azimuthal deflection in the RICH detectors.

In the first radiator, the magnetic field of the solenoid was negligible, while in the second radiator, external correction coils shaped the field to ensure that no Lorentz force acted there. The total material within the acceptance was minimized to less than 1% of a radiation length. Cherenkov photon detection in the UV range employed a

Fig. 3.38 Second generation heavy-ion experiment at CERN SPS: CERES/NA45. UV photon detectors were based on the photo effect in TMAE vapor at 40 °C and the whole spectrometer was heated to 50 °C. © University of Heidelberg. All right reserved

three-stage amplification system, consisting of two parallel-plate layers and one wire plane. The detectors were positioned upstream of the target to reduce background. The readout was done through electrodes with $2 \times 50\,000$ pads. For the Pb-beam era, silicon drift detectors and an additional pad chamber were added to enhance performance.

Historical Context and Collaboration

Initially, the CERES collaboration consisted of just three partners: MPI Heidelberg (P. Wurm et al.), the Weizmann Institute Rehovot (A. Breskin, I. Tserruya, et al.), and the Physics Institute Heidelberg, starting with only 22 members and growing to about 50. The subsequent influx of graduate and doctoral students during this period exceeded all my previous experiences in both quality (including several scholarship holders) and quantity (Fig. 3.39).

R&D and Running Timeline

The R&D program for the UV detectors spanned several years and resulted in numerous publications, including one co-authored with G. Charpak. The R&D phase was a dynamic and fast-paced period, leading to rapid construction of this pioneering experiment between 1989 and 1991. Two major running periods followed, during which the R&D program continued alongside significant upgrades:

– 1992–1993: Experiments with ^{32}S and proton beams (p-Be and p-Au), based on puristic hadron-blind tracking with two RICH detectors.
– 1995–1996: Experiments with ^{208}Pb beams, adding two SiDC detectors and a pad (multi-wire) chamber.

Fig. 3.39 CERN staff (J. Schukraft), Heidelberg (A. Drees, P. Fischer, A. Pfeiffer, C. Schwick, T. Ullrich, Hans J. Specht), and the Weizmann Institute (A. Breskin, I. Tserruya) at the CERES setup in 1991. These researchers from Heidelberg also continued their commitment to scientific research in the years that followed. © Hans J. Specht. All rights reserved

The early R&D and construction phase saw both smooth progress and unexpected challenges. In August 1990, the "CERES first-ring party" was held to celebrate the successful installation of CERES and the first observation of RICH rings. Despite this success, we soon faced the infamous "sparks problem"—a critical issue which presented a significant challenge but also motivated us to develop a robust solution (Figs. 3.40 and 3.41).

Resolving the Spark Problem

The CERES UV detectors were originally conceived as parallel-plate counters with two-step amplification, an intermediate gate-electron pair, and a final drift stage towards a pad electrode. However, when operated in beams of a few 10^6/burst protons or ^{32}S-nuclei at 200 GeV/u, the scheme suffered from excessive spark rates. The origin of the sparks was explained quantitatively by slow event-correlated secondary particles, which created avalanches above the critical threshold of about 10^8 charges at the required gain of a few 10^5. This lack of sufficient dynamic range severely limits the use of large-area parallel-plate counters for single-electron detection in a realistic high-energy physics environment (Fig. 3.42).

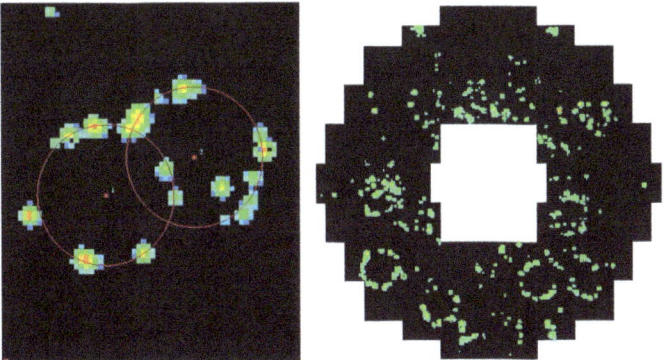

Fig. 3.40 RICH Cherenkov rings. © University of Heidelberg. All right reserved

Fig. 3.41 R&D and construction activities of CERES at the Physikalisches Institut in Heidelberg. © Hans J. Specht. All rights reserved

The spark problem of the CERES UV-detectors was solved by introducing wire amplification at the last stage. This breakthrough inspired us to publish our findings in a paper titled "*In-beam experience from CERES UV detectors: Prohibitive spark breakdown in multi-step parallel-plate chambers as compared to wire chambers*" (published in Nuclear Instruments and Methods in Physics Research A 343 (1994) 231–240).

Fig. 3.42 The components of the CERES detectors, constructed at the Physikalisches Institut of the University of Heidelberg, being transported to CERN for installation. © Hans J. Specht. All rights reserved

Reflecting on this achievement, the paper concluded: "We summarized our own personal view after completion of a full circle and two years of happy data taking in one short remark: we will in the future stick to the lovely tolerance of an MWPC."

Dilepton Excess: Revealing the First Hints of a New State of Matter

The first production beam time with 200 GeV/u S-Au collisions took place in 1992. The data revealed a strong excess of dilepton pairs, beyond what could be expected from known hadron decays, marking the first clear sign of new physics in the mass region below 1 GeV. The CERES e^+e^- mass spectrum measured in S-Au collisions is shown in Fig. 3.44 together with the measured CERES reference spectra for 450 GeV p-Be and p-Au collisions. While the p-Au data are reproduced within errors by final state Dalitz and direct decays of neutral mesons, as known from pp collisions, electron pairs from S-Au collisions reveal a substantial enhancement in the mass range above 250 MeV (Fig. 3.43).

The excess has to originate from processes which are active during the lifetime of the fireball, i.e., between the onset of hadronisation and kinetic freezeout. At this top SPS energy and close to the critical temperature, the prime candidate for "thermal radiation" from the hadronic phase of the fireball is pion annihilation. The observed excess of dileptons was consistent with the expectation of ρ-meson regeneration (lifetime of only 1.3 fm/c) during the fireball expansion through $\pi^+\pi^- \leftrightarrow \rho$ processes, with a dilepton branching of 1 in 10^4. This regeneration resulted in approximately

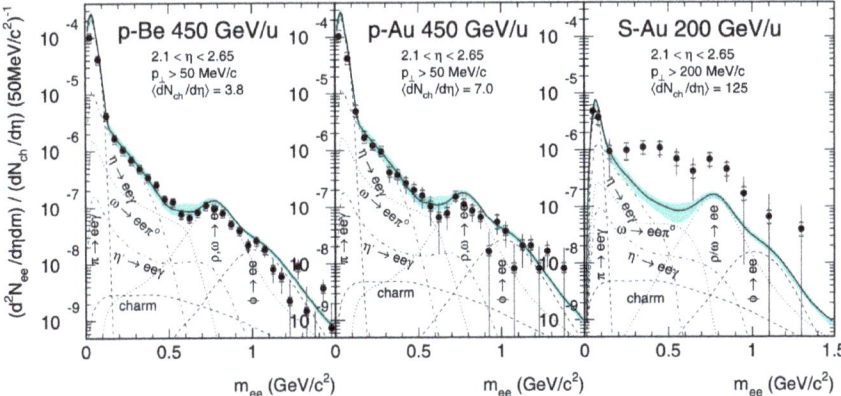

Fig. 3.43 CERES inclusive e^+e^- mass spectra of 450 GeV p-Be, p-Au, and 200 GeV/u S–Au collisions. Contributions from various hadron decays as expected from p-p collisions are shown together with their sum (thick line). © 2005 Springer. All rights reserved

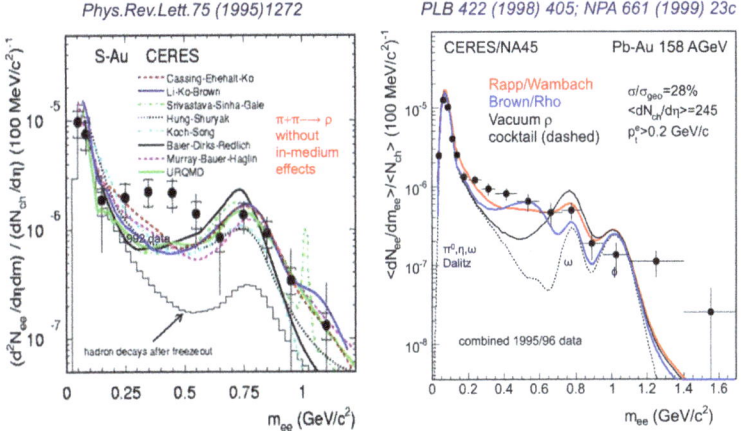

Fig. 3.44 Comparison of CERES results of dilepton invariant mass spectra for S–Au and Pb–Au collisions with different theoretical modeling. Strong excess of dileptons above meson decays, but mass shift and broadening of ρ still indistinguishable. © 1995 American Physical Society. All rights reserved (left panel) and © 2005 Springer. All rights reserved (right panel)

five times the normal yield in the long-lived hadronic phase of the fireball after the QCD phase transition.

Numerous theoretical approaches incorporating pion annihilation using vacuum properties of the ρ meson failed, without exception, to match the data. The observed bulk excess could only be described by invoking a strong in-medium modification of the ρ meson's spectral function. These modifications included a mass shift—directly linked to the restoration of chiral symmetry, which ties hadron masses to the value of the chiral condensate (the scaling conjecture of Brown and Rho)—or broadening, based on a hadronic many-body approach (e.g., the Rapp and Wambach calculation). However, the statistical accuracy of the data was not sufficient to distinguish clearly between these competing theoretical scenarios. Consequently, the question of how the ρ meson's spectral function changes in the medium near chiral phase transition, i.e., whether it shifts in mass or broadens, remained open.

The lead beam running period took place in 1995 and 1996, during which we measured dilepton production in 158 GeV/u Pb-Au collisions with a significantly improved setup compared to the sulfur beam experiment, leading to better mass resolution and statistical accuracy. The enhancement of the measured yield over the reference of neutral meson decays was again observed. The measured yield, its stronger-than-linear scaling with N_{ch}, and the dominance of low-pair p_T strongly suggest an interpretation as *thermal radiation* from pion annihilation in the hadronic fireball. To describe the shape of the excess, strong in-medium modification of the ρ-meson was again required. However, the ambivalent situation—mass shift or broadening—persisted into the Pb-beam era, as illustrated in Fig. 3.44 (right) for the CERES 1995/96 data (also valid for the 2000 data). This was finally resolved, in an outstanding way, in my last CERN experiment, NA60.

Nevertheless, these first publications of the results [*PRL 75 (1995) 1271, Ph.D. thesis by T. Ullrich and PLB 422 (1998) 405; NPA 661 (1999) 23c; Eur. Phys. J C 41 (2005) 475-513*] provided an enormous boost to theory with hundreds of further publications. To this day, the S-Au data remains not only the most cited work of my career in physics but also the most cited experimental paper in heavy-ion physics at the SPS. In 2000, at a historic CERN press conference, with RHIC on the horizon, the contribution made by CERES was highlighted as one of the key pieces of evidence supporting the official announcement "New State of Matter created at CERN" (Fig. 3.45).

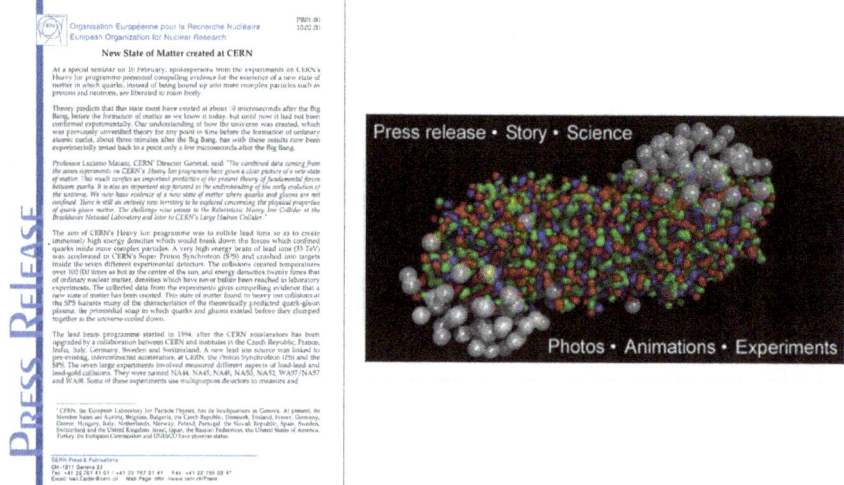

Fig. 3.45 Press Conference entitled 'New State of Matter Created at CERN,' 10 February 2000. © CERN, All rights reserved

Quark Matter Conference '96—Held in Heidelberg

The 1996 Quark Matter Conference in Heidelberg remains one of the most significant and moving events of my career. Co-organized by GSI (during my tenure as director) and the University of Heidelberg, the conference brought together about 560 scientists from around the world setting a new attendance record for the quark matter series (Figs. 3.46 and 3.47).

In over fifty presentations, the conference offered the first comprehensive overview of results from truly heavy-ion collisions, using the 160 GeV/nucleon Pb beam from CERN's SPS and the 11 GeV/nucleon Au beam from Brookhaven's Alternating Gradient Synchrotron (AGS).

Our ambitions were high: to present the very first CERES results from the Pb production run, completed just 5 months earlier. The timeline was tight, and even during those intense months, we undertook a major technical conversion of the software. Yet, we worked tirelessly, and I will always remember how, just 5 days before the conference, we were finally ready to unveil the results. The collective effort, passionate involvement, and determination of the team culminated in the presentation of the CERES results on the dilepton invariant mass spectra for Pb-Au collisions, which became one of the conference highlights. It was a moment which reflected years of hard work, making Quark Matter '96 not just a scientific milestone but also a deeply personal and professional triumph. Seeing the CERES results resonate with the community remains one of the most rewarding moments of my career.

The review from the QM96 Conference titled "Report of a Little Bang" was prepared by Helmut Satz and published in the CERN Courier, July/August 1996. As Satz aptly noted: "Heidelberg 1996 could be remembered as the first report of a little bang."

Fig. 3.46 The international community convenes: Quark Matter '96, Heidelberg, May 1996. © Hans J. Specht, All rights reserved

Fig. 3.47 Hans J. Specht at the opening ceremony of the QM 96 Conference. © Hans J. Specht, All rights reserved

A Special Musical Composition for Quark Matter '96 in Heidelberg

Beyond the groundbreaking scientific presentations, the Quark Matter '96 conference had an extraordinary cultural dimension, highlighted by both a jazz evening and a memorable classical concert. The latter stood out as one of the event's most unforgettable moments, particularly because of a special gift presented to commemorate the occasion: the talented composer Professor Martin Messmer created a unique composition titled *Phase/Plasma 2, 1996*.

Martin Messmer, Professor of Music Theory at the Hochschule für Musik und Theater in Hannover, has collaborated with prestigious ensembles and artists, including the Balanescu Quartet London, John Kenny, Robyn Schulkowsky, Ensemble Recherche Freiburg, Ex Voco, and Süddeutscher Rundfunk, among others. The piece was performed for the first time at the QM96 Classical Concert in Alte Aula and was specially dedicated to the Quark Matter Conference. This profoundly moving gesture left a lasting impression on me.

Martin Messmer's explanation of the inspiration behind *Phase/Plasma 2*, which I fondly recall, was included in the Programme Notes and is shared here:

"The composition Phase/Plasma 2 was written in March and April 1996, at the suggestion of Prof. Hans J. Specht. It will be premiered on 21 May 1996 on the occasion of the conference Quark Matter 96 in Heidelberg.

"Phase/Plasma 2 has been written for oboe, clarinet, violin, viola, cello, double bass and piano and is the sequel to an earlier work, Plasma, composed in 1993. Conversations about and discussions on quantum physics and elementary particle physics were the influences behind the formation process of the composition on both a technical and an aesthetic level.

"Two different states of subjectively experienced statics oppose each other as dialectical poles. A small, very fast-moving sound surface is perceived in a similar vein to barely moving, fixed objects. The absence of simple melodies, where the relationships between the velocities (rhythm) appear in conventional quantities and where movement only exists in extremely high or very low velocities, demands another form of perception. A sound surface created by several rhythmic planes superimposed upon one another seems chaotic (which, in fact, it is; the individual sound events in such structural spheres, spaced horizontally along an imaginary melodic line, could only be realized in absolute perfection by a computer). As this sound surface is created by interweaving previously-defined single lines, the composer also has to take the (albeit very slight) influence of chance into account.

"As previously mentioned, melodic developments in such a field cannot be perceived; rather, through the gradual change of field density, complexity, and velocity, a very slow, almost virtual time structure sets in, which in this way corresponds with the parts of actual statics. To do this, I have made use of special forms of the techniques of so-called minimal music.

"I have tried to obtain a transmission of the fluctuating and rhythmically almost indistinct principle of the moving static parts onto the linear (horizontal) plane by quarter-tone scordatura of the strings (the stringed instruments are not used with their normal tuning; the individual strings have to be "untuned" in a specifically defined way). Microtonal impurities are thus formed, in which, for example, linear fragments are constantly surrounded by slightly "out," "wrong" notes. It thus becomes impossible to perceive the exact frequency of the note, yet it is possible that roughness effects and beat frequencies may be felt.

"To be able to realise these sound effects in a precise manner, it was necessary to eliminate all subjective, that is, human, components (such as slightly inaccurate playing, vibrato, etc.). Consequently, the strings only play open strings and harmonics."

Transitioning to New Horizons

In 1996, P. Braun-Munzinger and J. Stachel were appointed to Darmstadt and Heidelberg, respectively, appointments in which I, as GSI Scientific Managing Director at the time, played a significant role. At my invitation, both joined CERES to bridge the long preparation period until ALICE was operational at the LHC. This collaboration, with J. Stachel as the new spokesperson, led to the addition of a radial TPC, allowing for hadron physics to be incorporated into the experiment. The collaboration saw substantial growth with the addition of new members, while my own group, due to my absence in Heidelberg, dwindled almost to zero. New beam times with Pb beams were first conducted in 1999 at an SPS energy of 40 GeV/u, followed by a full-energy run in 2000. Apart from a new Ph.D. student analyzing electron pair data at 40 GeV/u—which also became a highly-cited success with an even larger effect in the ρ region—I found myself quite isolated. As CERN did not approve additional CERES beam times at lower energies, it became clear that the time was ripe for a new adventure, especially since the ongoing status of lepton pair research, in mass regions both below and above 1 GeV, still did not meet the high expectations of the goals set during the Quark Matter 2 workshop in Bielefeld two decades earlier.

Experiment NA60 at SPS

Solid Proof of Deconfinement at SPS Energies

Once again, fortune was on my side, this time on a level which surpassed all my previous experiences at CERN. Two senior colleagues from the NA50 muon pair experiment, L. Kluberg and P. Sonderegger, along with C. Lourenço, who served as spokesperson for the successor experiment NA60 until 2005, had long been encouraging me to join NA60. Given my expertise with low-mass lepton pairs, which complemented the focus of NA50 on higher mass ranges above 1 GeV and J/ψ physics, they saw it as a perfect match. I accepted the offer in 2003, just before my retirement in 2004 and the impending end of federal funding. Fortunately, early successes postponed these funding cuts until 2006, allowing me to make substantial contributions to both hardware and, more crucially, simulation efforts, working with a dedicated team of postdocs. Together, we built high expectations leading up to the first beam time at the end of 2003.

The significant advancement from NA50 to NA60 can be credited to a truly ingenious idea by P. Sonderegger: the integration of a 0.5-m-long precision spectrometer with a novel radiation-hard silicon pixel vertex tracker, embedded in a 2.5 T dipole magnet. This setup was positioned in the target region, ahead of the 5-m-long hadron absorber, and followed by the traditional NA50 spectrometer. Track matching between two spectrometers, both in coordinate and momentum space, greatly improved the dimuon mass resolution by a factor of 4 compared to NA50, allowing for a clear distinction between prompt and decay dimuons (from open charm). Additionally, the dipole field at the beginning of the setup improved the overall acceptance at low p_T. The silicon detectors used in NA60 represented a groundbreaking development, originally designed for LHC experiments. Their implementation in NA60 marked the first use of such detectors in a running experiment of the version with high tolerance to radiation damage. The radiation hardness of the Si tracker, together with a very high readout speed, enabled the maintenance of high luminosities typical for muon pair measurements, achieving a factor of 1000 greater than those attainable with CERES. Additionally, the muon pair trigger further contributed to this achievement. By taking a big step forward in technology, this third-generation experiment achieved completely new standards of data quality in the field (Figs. 3.48 and 3.49).

3 Hans Joachim Specht (in His Own Words) …

Fig. 3.48 NA60 experiment at CERN SPS. © 2003–2004 CERN, All rights reserved

NA60 Dilepton Excess Unveiled: High-Quality Data Enables Model-Independent Analysis

For the first time, the high data quality could be used to isolate dilepton excess without any assumptions about its nature and reliance on fits.

Figure 3.50 illustrates the NA60 data sample and its quality in the low-mass region. After subtracting the total background, the resulting net spectrum contains 400 000 dimuons in this mass region. This represents an improvement of up to three orders of magnitude in effective statistics and a factor of 2–5 in mass resolution compared to the previous results. The vector mesons ω and ϕ are fully resolved, and even the two-body η decay is observed, with the mass resolution at ω being 20 MeV.

Ending a Decades-Long Controversy over Chiral Symmetry Restoration

In the low-mass region, the isolation of the excess reveals the ρ spectral function. The cocktail of decay sources is subtracted from the total data using local criteria, based solely on the measured mass distribution itself, with the ρ meson intentionally

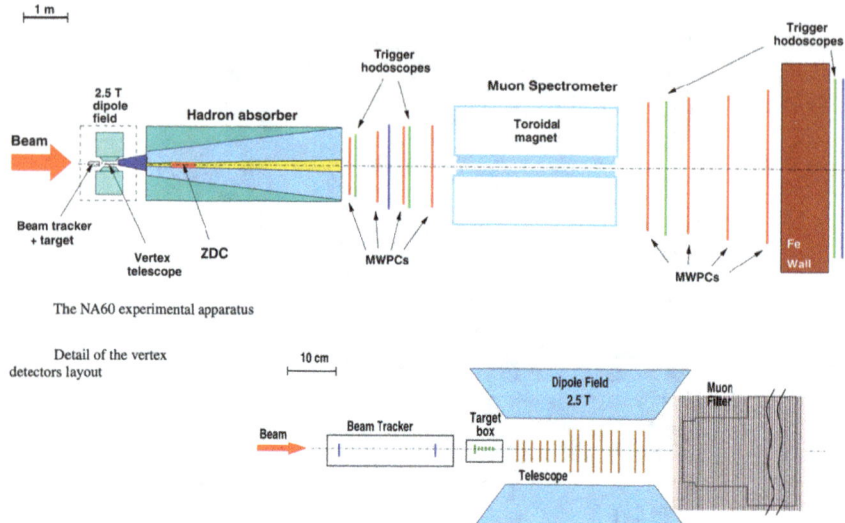

Fig. 3.49 Setup of the NA60 experiment at CERN SPS. © 2009 Springer. All rights reserved

Fig. 3.50 High-precision measurement of the dilepton invariant mass spectrum by the NA60 experiment. © 2006 American Physical Society. All rights reserved

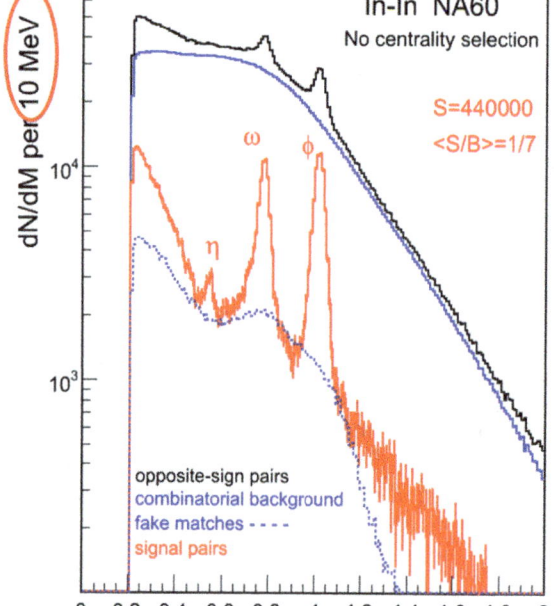

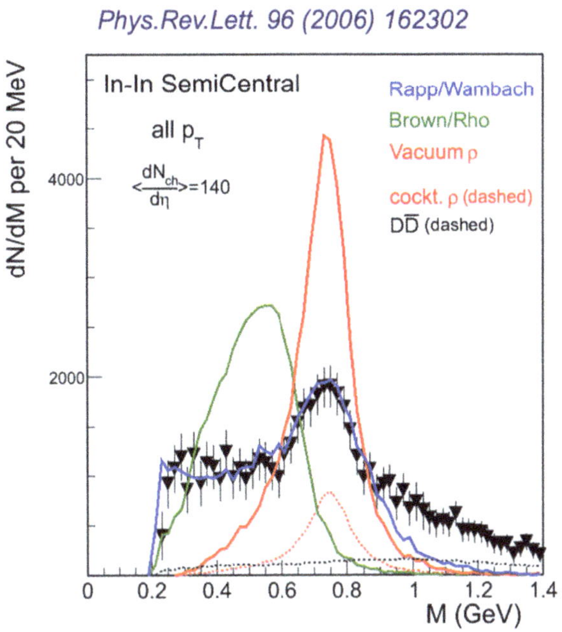

Fig. 3.51 Space–time averaged ρ spectral function. Comparison of the NA60 excess data with theoretical model results. © 2006 American Physical Society. All rights reserved

not subtracted. Figure 3.51 shows the excess in the mass range below 1 GeV, where the ρ meson appears as a peak at its nominal position. The "in-medium" effects manifest themselves solely as a broadening—described as the "melting" of the ρ meson (explained theoretically by van Hees and Rapp)—with no shift in its mass. Thus, after more than 20 years, this clearly rules out the "Brown–Rho scaling" hypothesis, which directly ties hadron masses to the value of the chiral condensate (with vanishing values as chiral restoration is approached) and predicts a mass shift of the ρ meson. The NA60 results, measuring the space-time averaged ρ spectral function, conclusively ended a decades-long controversy about the spectral properties of hadrons close to the QCD phase boundary.

A True Thermal Planck-Like Spectrum

In the absence of resonances, the signature of any thermal source should be a Planck-like radiation spectrum. With NA60's measurement of such a spectrum in high-energy nuclear collisions, isolated from all other sources, a 25-year-old dream became reality.

Figure 3.52 shows the final result for the thermal dimuon mass spectrum across the entire range of $0.2 < M < 2.5$ GeV. The combinatorial background and all known hadronic decay contributions, including Drell–Yan and open charm (measured by

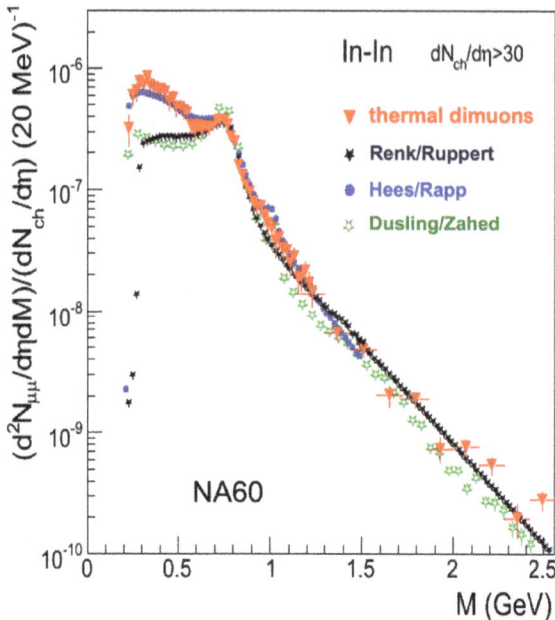

Fig. 3.52 Thermal muon pair mass spectrum in In-In collisions. © CERN, CERN Courier 49N9 (2009) 31–34. All rights reserved and © 2009 Springer. All rights reserved

displayed decay vertices), but excluding the ρ, have been meticulously subtracted. The data are integrated over p_T, acceptance-corrected, and normalized to the central rapidity density of charged hadrons, allowing for direct comparison with theoretical calculations.

The shape of the spectrum predominantly follows a pure exponential, indicative of a flat spectral function as in the black-body case, except for the slight modulation around the nominal pole position of the ρ meson. The mass spectrum is precisely described by $dN/dM \propto M^{3/2} exp(-M/T)$ [Shuryak 1978], where T represents the average temperature over the space-time evolution of the system. Since M is Lorentz invariant, the spectrum is not affected by the expansion of the fireball (which ends at $c/2$), unlike the Planck distribution for photons. Thus, T reflects the purely thermal temperature of the system. Fitting the data with this Lorentz-invariant "Planck-like" function above 1.1 GeV yields a temperature value of 215 ± 12 MeV. Given the critical temperature of 160–170 MeV for the QCD phase transition, this indicates that partonic sources dominate the observed thermal radiation.

This provides the clearest evidence to date, derived directly from the data, without relying on model calculations, that the quark-gluon plasma was indeed first created and observed at CERN SPS. This evidence is more direct and clear cut than the prior findings available before the CERN press conference in 2000, which necessitated cautious language at that time.

Reflection on the Search for Thermal Dileptons: From the Goals in 1980s to Achieving a Clear Signal

For me personally, it is more than satisfying to see that the goal set in the early 1980s, namely to make clear statements in such a complex field, has finally been achieved.

The dimuon production data collected by NA60 are exceptionally rich, featuring the explicitly projected ρ spectral function, m_T spectra, radial expansion and transverse flow of the thermal radiation, helicity distributions, form factors, and more. This wealth of data resulted in a series of publications. A comprehensive overview of all NA60 results on thermal radiation in the form of muon pairs can be found in CERN Courier, 11 (2009) 31, authored by S. Damjanovic, R. Shahoyan, and Hans J. Specht. G. Usai has been the spokesperson for NA60 since 2005.

Selected Publications—A Guide

High-Energy Physics: towards Heavy Ions at CERN ("Quark Matter")

- Future Experiments (Summary Talk)

  H.J. Specht

  in Proc. Workshop on Future Relativistic Heavy Ion Experiments

  Darmstadt 1980, GSI 81-6; Darmstadt (1981), 551–558

  The workshop at GSI unofficially named "Quark Matter I." This was a public plea, in the spirit of the meeting, to use CERN accelerators for quark matter research as opposed to a non-competitive machine at GSI (SIS100).

- Inclusive Measurements and Particle Identification

  S. Nagamiya and H.J. Specht

  Workshop on Quark Matter Formation and Heavy-Ion Collisions,

  Bielefeld 1982, 519–536; eds. M. Jacob and H. Satz, World Scientific.

- Direct Photons and Lepton Pair Production in High Energy Collisions

  H.J. Specht

  Quark Matter 1984, Proc. of the 4th International Conference, Helsinki,

  Lecture Notes in Physics 221 (1984) 221–239, Springer.

- Production of Prompt Positrons at low p_T in 63 GeV pp Collisions at the ISR

  T. Akesson et al., The Axial Field Spectrometer (AFS) Collaboration

Phys. Lett. 152B (1985) 411–418; Phys. Lett. B192 (1987) 463–470

The only hints of an excess of low-mass dileptons in high-energy pp so far.

- Soft Photon Production in 450 GeV/c p-Be Collisions

    J. Antos, H. Beker, S. Brons, K. Bussmann, S. Dagan, A. Drees, C. Erd, M.J. Esten, C.W. Fabjan, P.Glässel, U.Görlach, V. Hedberg, D. Lissauer, M.A. Mazzoni, N.A. McCubbin, M. Neubert, P. Nevski, L. Olsen, A. Pfeiffer, A. Ray, J. Schukraft, D. Shapira, J. Soltani, H.J. Specht, I. Stumer, J. Thompson, R.J. Veenhof, W.J. Willis, and C. Woody

    Z. Phys. C59 (1993) 547–553

    A mysterious photon source with $pT < 20$ MeV, most recently also observed in the most forward region of jets (in DELPHI).

- Low-Mass Lepton-Pair Production in p-Be Collisions at 450 GeV/c

    T. Akesson et al., HELIOS Collaboration

    Z. Physik C 68 (1995) 47–64

    Dielectrons and dimuons from the same experiment. In contrast to previous experiments, there was no evidence for low-mass "anomalous" pair production in moderate-energy pp. Soon after, this was also confirmed by NA45/CERES for pBe and pAu.

Pre-ALICE Discussion

- Experimental Aspects of Heavy Ion Physics at LHC Energies

    H.J. Specht

    Proc. ECFA Large Hadron Collider Workshop, eds. G. Jarlskog and D. Rein,

    CERN 90–10 (1990) Vol. II, 1236–1251

    Convener of the experimental discussion on a future LHC heavy-ion program at the Aachen Workshop, providing an early version of an LHC heavy-ion detector as input to the foundation phase of ALICE one year later.

Heavy-Ion Results/NA34-2-HELIOS/S-beams only

- Inclusive negative particle p_T spectra in p–nucleus and nucleus–nucleus collisions at 200 GeV/u

    T. Akesson, HELIOS Collaboration

    Z. Phys. C46 (1990) 361–367

    Central results from the external spectrometer (Ph.D. thesis A. Drees).

- Inclusive photon production in pA and AA collisions at 200 GeV/u

  T. Akesson et al., HELIOS Collaboration

  Z. Phys. C46 (1990) 369–375

  No significant excess of photons above the known hadron decay sources.

- Measurement of the Transverse Energy Flow in Nucleus-Nucleus Collisions at 200 GeV per Nucleon

  T. Akesson et al., HELIOS Collaboration

  Nucl. Phys. B353 (1991) 1–19

  Most accurate measurement of the energy flow at the SPS, thanks to the use of a 4π calorimeter (recycled from R807/808 at the ISR). Targets up to U, showing a softer slope at the kinematic limit than all others due to nuclear deformation ("the most expensive way to measure nuclear deformation").

Heavy-Ion Results / NA45-CERES / Electron Pairs

- A Highly Efficient Low-Pressure UV-RICH Detector with Optical Avalanche Recording

  A. Breskin, R. Chechik, Z. Fränkel, D. Sauvage, V. Steiner, I. Tserruya, G. Charpak, W. Dominik, J.P. Fabre, J. Gaudean, F. Sauli, M. Suzuki, P. Fischer, P. Glässel, H. Ries, A. Schön, and H.J. Specht

  Nucl. Instr. Meth. A273 (1988) 798–802

  R&D: potentially interesting, but a dead end due to practical limitations.

- In-Beam Experience from the CERES UV-Detectors: Prohibitive Spark Breakdown in Multi-Step Parallel-Plate Chambers as Compared to Wire Chambers

  R. Baur, A. Drees, P. Fischer, Z. Fränkel, P. Glässel, H. Klein, A. Pfeiffer, A. Schön, A. Shor, H.J. Specht, V. Steiner, I. Tserruya, Th. S. Ullrich

  Nucl. Instr. Meth. A343 (1994) 231–240

  R&D: a very personal paper, with regard to both the decisive data on saturation of wire amplification and the spirit of the text, reflecting the frustration of the year 1991, with CERES on beam but unusable due to prohibitive sparking.

- The CERES RICH Detector System

  R. Baur, A. Breskin, R. Chechik, A. Drees, U. Faschingbauer, P. Fischer, Z. Fränkel, J. Gläss, P. Glässel, C.P. de los Heros, D. Irmscher, R. Männer, A. Pfeiffer, A. Schön, J. Schukraft, Ch. Schwick, A. Shor, H.J. Specht, V. Steiner, S. Tapprogge, G. Tel-Zur, I. Tserruya, Th. Ullrich, and J.P. Wurm

Nucl. Instr. Meth. A343 (1994) 87–98

Successful R&D: the final CERES RICHes have reached the theoretical limits for the number of UV photons/ring and the spatial resolution of the ring centers, determining the mass resolution of the whole spectrometer.

- Enhanced Production of Low-Mass Electron-Pairs in 200 GeV/u S–Au Collisions at the CERN Super Proton Synchrotron

  G. Agakichiev, R. Baur, A. Breskin, R. Chechik, A. Drees, C. Jacob, U. Faschingbauer, P. Fischer, Z. Fraenkel, Ch. Fuchs, E. Gatti, P. Glässel, Th. Günzel, C.P. de los Heros, F. Hess, D. Irmscher, B. Lenkeit, L.H. Olsen, Y. Panebrattsev, A. Pfeiffer, I. Ravinovich, P. Rehak, A. Schön, J. Schukraft, M. Sampietro, S. Shimansky, A. Shor, H.J. Specht, V. Steiner, S. Tapprogge, G. Tel-Zur, I. Tserruya, Th. Ullrich, J.P. Wurm, and V. Yurevich

  Phys. Rev. Letters 75 (1995) 1272–1275

  The first clear sign of new physics from dileptons: strong excess yield above the known meson decays, attributed to thermal radiation mediated by the ρ meson. Most cited data paper (>630) of the whole SPS ion program (Ph.D. thesis T. Ullrich). Original CERES setup.

- e^+e^- pair production in Pb-Au collisions at 158 GeV per nucleon.

  G. Agakichiev et al., CERES Collaboration

  Eur. Phys. J. C 43 (2005) 475–513

  The final summary paper of all Pb-Au results obtained with the original CERES set-up: on mass spectra, p_T spectra and multiplicity dependences. Better resolution than in S–Au; confirmation of the strong excess yield above the known sources, but the mass shift and broadening of the ρ remained indistinguishable due to insufficient statistics. For the same reason, no insight was gained into the dilepton mass region >1 GeV (1995 data Ph.D. thesis C. Vogt, 1996 data Ph.D. thesis B. Lenkeit).

- Enhanced Production of Low-Mass Electron–Positron Pairs in 40 A GeV Pb-Au Collisions at the CERN SPS

  D. Adamova et al., CERES Collaboration

  Phys. Rev. Lett. 91 (2003) 042301/1-5, nucl-ex/0209024 (2003)

  So far, the only dilepton results in the lower energy region close to the onset of the expected QCD phase transitions (Ph.D. thesis S. Damjanovic). Very large dilepton excess. CERES TPC setup.

- Experimental Conference Summary

  Hans J. Specht

  Proc. Quark Matter 2001, Stony Brook, Nucl. Phys. A698 (2002) 341c–359c

  A unique moment in the history of the field: the results of the SPS program had just been presented in a CERN Press Release in early 2000 as a "New State of Matter created at CERN," and the first set of results after the start-up of RHIC later in 2000 had already been analysed with an admirable efficiency, creating an atmosphere of enthusiasm and allowing for immediate comparisons.

CERN LHC General Interest

- Study of potentially dangerous events during heavy-ion collisions at the LHC

  J.P. Blaizot, J.Iliopoulos (Chair of the Ad-Hoc Committee), J. Madson, G.G.Ross, P. Sonderegger, and H.J. Specht

  Yellow Report CERN 2003-001

  A review of dangerous objects such as negatively charged strangelets, gravitational black holes, and magnetic monopoles. Conclusion: "We find no basis for any conceivable threat."

Heavy-Ion Results—NA60/Muon Pairs

- First Measurement of the ρ Spectral Function in High-Energy Nuclear Collisions

  R. Arnaldi et al., NA60 Collaboration

  Phys. Rev. Letters 96 (2006) 16302

- Evidence for Radial Flow of Thermal Dileptons in High-Energy Nuclear Collisions

  R. Arnaldi et al., NA60 Collaboration,

  Phys. Rev. Lett. 100 (2008) 022302

- First Results on Angular Distributions of Thermal Dileptons in Nuclear Collisions

  R. Arnaldi et al., NA60 Collaboration

  Phys. Rev. Lett. 102 (2009) 222301

- NA60 Results on Thermal Dimuons

  R. Arnaldi et al., NA60 Collaboration

  Eur. Phys. J C 61 (2009) 711

- Evidence for the Production of Thermal Muon Pairs with Masses above 1 GeV in 158A GeV Indium–Indium Collisions

  R. Arnaldi et al., NA60 Collaboration

  Eur. Phys. J. C 59 (2009) 607

  Superior quality of the data compared to all other experiments in this field: effective statistics higher by a factor of nearly 1000, unattainable by any collider in the future. Another factor of 100 attainable in a future "NA60+", under development for an energy scan of 20–160 AGeV at the CERN SPS.

  Main conclusions: (i) close to the QCD transition, the in-medium ρ spectral function broadens ("melts"), but shows no shift in mass; (ii) above masses of 1 GeV, i.e., in the region of flat spectral functions, both the slope $T > Tc$ of the dilepton invariant mass spectrum and the very low radial flow indicate the dominance of a partonic emission source in this region, a clear signal of deconfinement at SPS energies; (iii) zero polarization of the radiation, consistent with a thermalized system.

- Thermal Dileptons from Hot and Dense Strongly Interacting Matter

  Hans J. Specht, NA60 Collaboration

  AIP Conf. Proceedings 1322 (2010) 1–10, nucl-ex 1011.0615 (2010)

  An up-to-date summary of the central NA60 results on thermal dileptons, also emphasizing the model-independence of the average temperature extracted from the high-mass part of the mass spectrum ("the only Lorentz-invariant thermometer in the field").

Hadron Structure/NA60

- Study of the electromagnetic transition form-factors in $\eta \to \mu^+\mu^-\gamma$ and $\omega \to \mu^+\mu^-\pi^0$ decays with NA60

  R. Arnaldi et al., NA60 Collaboration

  Phys. Lett. B 677 (2009) 260

- Precision study of the $\eta \to \mu^+\mu^-\gamma$ and $\omega \to \mu^+\mu^-\pi^0$ electromagnetic transition form-factors and of the rho line shape in NA60

  R. Arnaldi et al., NA60 Collaboration

  Phys. Lett. B 757 (2016) 437–444

Confirmation of the long-standing strong violation of vector meson dominance (VMD) in the region close to the kinematic cutoff of the ω Dalitz decay as measured in the 1980s, with statistics improved by factors of 10 and 100, respectively. So far, theoretical models are unable to describe this. The results appeared in detail in the publications of the Particle Data Group (PDG) from 2010 and were at the time the first ever from a heavy-ion experiment.

Preparing the Future: Hans J. Specht and the Birth of ALICE at the LHC

Jurgen Schukraft

Building on his leadership in the SPS heavy-ion program, Hans J. Specht played a pivotal role in launching the LHC heavy-ion program, which had to be developed alongside preparations for the SPS Pb program (initiated in 1994) and the RHIC collider at BNL (which began in 2000). Together with several colleagues, Specht helped design an early concept for a dedicated heavy-ion detector at the LHC. His vision was first documented in the proceedings of the Large Hadron Collider Workshop, held in Aachen in October 1990 and organized by the European Committee for Future Accelerators (ECFA). These initial designs laid the foundation for what would evolve into ALICE, the third-generation heavy-ion experiment at CERN's LHC (Fig. 3.53).

The Early Design of a Heavy-Ion Detector

The conceptual origins of ALICE date back to the early 1990s, when CERN was considering the inclusion of heavy-ion beams at the LHC, reaching energies of up to 3.5 TeV per nucleon. To assess the physics potential of such collisions, ECFA established a specialized working group, *Heavy Ion Physics at the LHC*. Helmut Satz served as the theoretical convener, while Hans J. Specht was appointed as the experimental convener.

With the invitation message "*There is much interesting work to do,*" the group attracted leading scientists eager to explore both theoretical and experimental goals for heavy-ion research at the LHC. As experimental convener, Specht's primary task was to draft a comprehensive blueprint for a heavy-ion experiment, integrating key observables to probe the fundamental properties of nuclear matter at extreme energies.

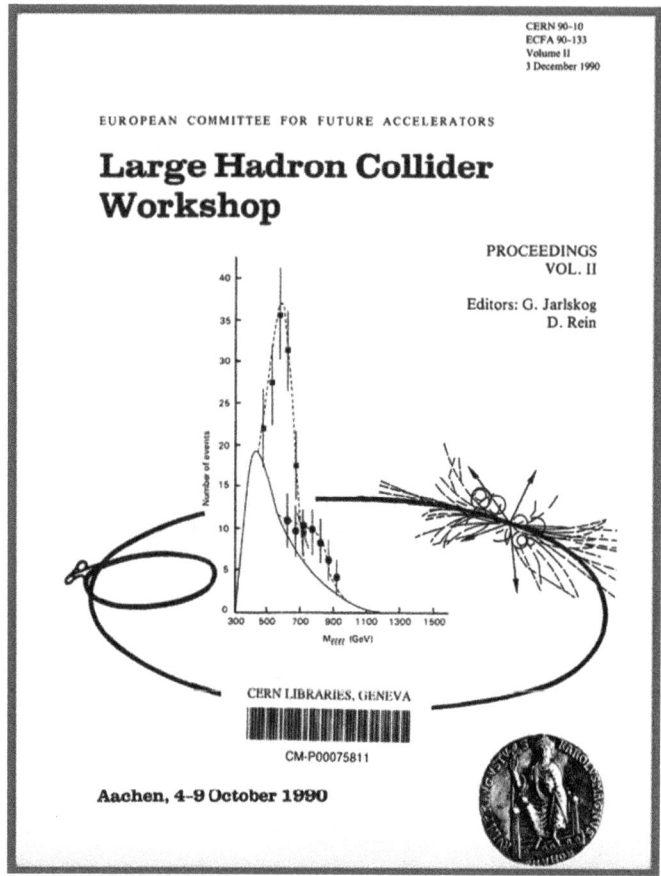

Fig. 3.53 Kick-off event: European Committee for Future Accelerators—Large Hardon Collider Workshop, Aachen, 4–9 October 1990. © CERN. All rights reserved

Designing such an experiment required balancing multiple factors—event rates, background suppression, and technical feasibility. Specht approached this challenge with his characteristic precision, developing two initial detector concepts, each carefully weighing scientific ambition against practical constraints.

Two Concepts for an LHC Heavy-Ion Detector

Although preliminary, Specht's early detector concepts demonstrated the technical sophistication required for a general-purpose heavy-ion experiment. Both designs minimized the amount of material near the interaction point to improve the signal-to-background ratio for electromagnetic probes—a physics topic especially close to Specht's heart (Fig. 3.54).

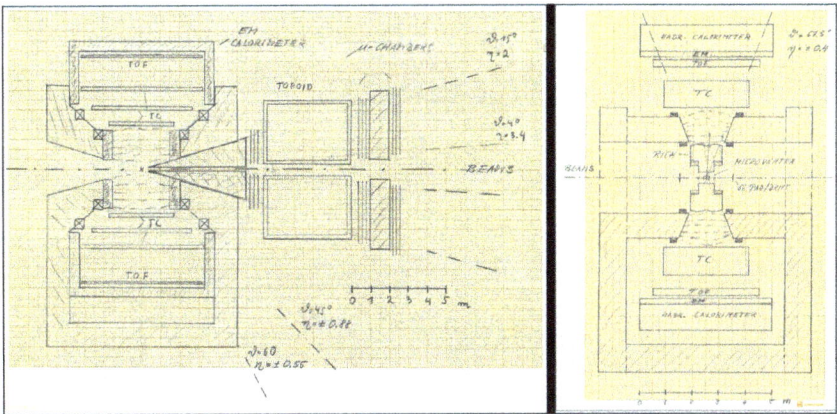

Fig. 3.54 Specht's meticulous hand-drawn designs of Concepts I and II for a universal heavy-ion detector at the LHC. © Hans J. Specht and © CERN. All rights reserved

- **Concept I**: This design featured an open, axially symmetric magnet with tracking chambers (TC) and a segmented time-of-flight (TOF) system for precise hadron measurements. Photons were detected using a high-resolution electromagnetic calorimeter, while a forward spectrometer captured muon pairs through an opening in the magnet yoke.
- **Concept II**: This alternative employed two pairs of Helmholtz coils to generate an axially symmetric magnetic field, with a zero-field region at the center. A microvertex detector was included for hyperon decay detection, while a ring-imaging Cherenkov (RICH) detector identified electron pairs. A high-resolution electromagnetic calorimeter measured photons, and silicon detectors in the forward region provided a multiplicity trigger. This design was optimized for low-mass electron pairs and incorporated the "zero material & zero field at the vertex" technique, similar to the CERES experiment, to suppress conversion electrons.

Published in the 1990 ECFA Aachen Workshop proceedings, these early designs demonstrated the feasibility of a single, general-purpose heavy-ion detector. They laid the groundwork for ALICE, and just two months later, in December 1990, Peter Sonderegger and Jürgen Schukraft organized the first meeting of interested heavy-ion physicists, aiming to form a dedicated LHC heavy-ion collaboration and propose a detector.

Specht's Role in the Formation of ALICE

The first official ALICE *Letter of Intent* (LoI) was published in 1993. While Specht facilitated the involvement of his Heidelberg research group in ALICE, he opted to serve in an advisory role on the Executive Committee rather than becoming a full

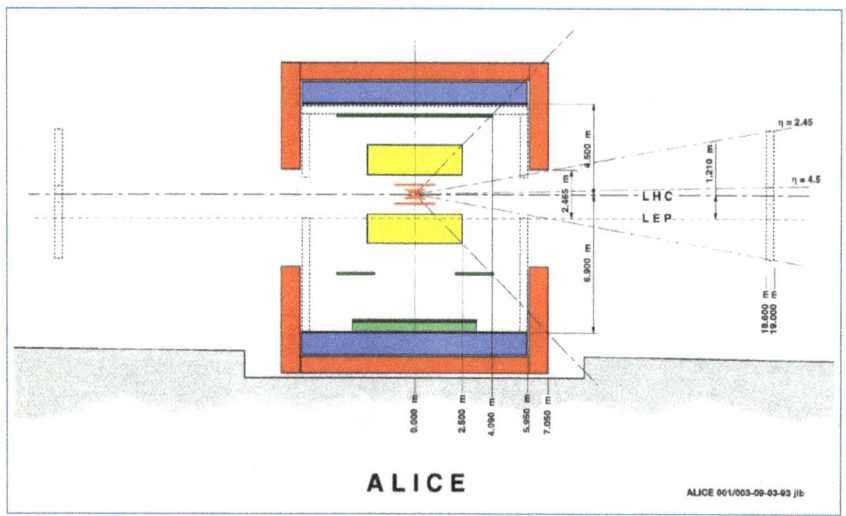

Fig. 3.55 From the first design for the Large Hadron Collider (LHC) heavy-ion detector to the ALICE detector (longitudinal view). LoI 1993. © CERN. All rights reserved

member of the collaboration. This decision reflected his concurrent commitments to other projects, including the CERES experiment at the SPS and his new position as Scientific Managing Director of GSI.

The LoI initially presented two very different layout options: a daring ultra-thin magnet positioned at the center of the detector acceptance, and a larger, more practical version based on the existing L3 magnet, although such a version was unavailable at the time. Largely because Specht insisted on "*focusing on the optimal but feasible,*" the first option was dropped, resulting in a more realistic and focused proposal—potentially instrumental in gaining approval for the experiment.

At GSI, Specht played a crucial role in forging a partnership between GSI and CERN. He secured GSI funding for ALICE and successfully advocated for German university groups to receive BMFT (Federal Ministry for Research and Technology) support to join the collaboration. GSI's contributions to ALICE included key detector components and the lead-ion injector for CERN's LINAC, ensuring a stable supply of heavy-ion beams for the experiment (Fig. 3.55).

While he ultimately shifted his focus to other projects, Hans J. Specht's foundational contributions were instrumental in transforming the idea of heavy-ion physics at the LHC into reality. His early conceptual work, leadership in the ECFA working group, and advocacy for institutional support were critical in shaping ALICE into one of CERN's flagship experiments.

Selected Publications

Pre-ALICE Discussion

- Experimental Aspects of Heavy Ion Physics at LHC Energies

  H.J. Specht

  Proc. ECFA Large Hadron Collider Workshop, eds.G. Jarlskog and D. Rein,

  CERN 90–10 (1990) Vol. II, 1236–1251

  https://cds.cern.ch/record/364098?ln=en.

First ALICE Official Document

- The first official ALICE document (ALICE Collaboration)—Letter of Intent, 1993, CERN: https://cds.cern.ch/record/290825?ln=en.

Reflections on the SPS Heavy-Ion Program: Past Achievements and Future Horizons

Sanja Damjanovic and Gianluca Usai

On the occasion of CERN's 60th anniversary and the celebration of 30 years since the first heavy-ion beam at CERN, held on 9 November 2016, Hans J. Specht delivered an inspiring talk in which he reflected on the birth of the ultra-relativistic heavy-ion program and the enduring legacy of the Super Proton Synchrotron (SPS). As the sole speaker summarizing the great physics results achieved by SPS experiments over three decades, Hans J. Specht provided a unique perspective on the evolution of these experiments and their discoveries (Fig. 3.56).

Heavy-Ion Physics at SPS and Hans J. Specht—A Long Journey Together

In his conclusion, Hans J. Specht highlighted the pivotal role of the SPS in the early exploration of quark–gluon plasma (QGP) formation and the broader QCD phase diagram. The experiments conducted at SPS enabled decisive learning processes regarding the relevant physical observables and the problem of matching them with appropriate experimental techniques, thereby guiding the development of next-generation experiments at RHIC and LHC.

Key achievements of the SPS heavy-ion program include:

- **Hydrodynamic-like time evolution of the collisions**
  Initial quark–gluon plasma (QGP) formation, followed by hadron formation in a state of "chemical" equilibrium at a temperature of $T\sim170$ MeV consistent with

Fig. 3.56 Celebration event on 9 November 2016 marking the 30th anniversary of the first heavy-ion beam at CERN. From right to left: Fabiola Gianotti (DG of CERN), Hans Gutbrod, Reinhard Stock, Hans J. Specht, and Urs Wiedemann (with Guy Paic in the background). Hans J. Specht, Reinhard Stock, and Hans Gutbrod were the spokespersons for the first-generation heavy-ion experiments NA34, NA35, and WA80, respectively, conducted between 1984 and 1987. © 2016 CERN, for the benefit of the ALICE Collaboration. All rights reserved

Hagedorn's statistical models, and "kinetic" freeze-out at T∼120 MeV after an "explosive" expansion reaching velocities near 0.5 c.

- **Evidence for quark–gluon plasma (QGP) formation**
  Key indicators included strangeness enhancement, suppression of J/ψ mesons (early evidence of color screening effects in the QGP), and hints of chiral symmetry restoration observed through in-medium modification of the ρ-meson spectral function. Decisive proof of the existence of the QGP came from the measurement of thermal dileptons, which directly determined the temperature of the early-stage fireball at an average of approximately $T \sim 220$ MeV—well above the critical temperature (T_c) for deconfinement.

A Vision for the Future Role of the SPS

Hans J. Specht's vision for the SPS extended far beyond its past achievements. He firmly believed that the SPS was far from being a machine of the past, but it remains a vital tool for addressing some of the unresolved questions in nuclear physics, particularly in the study of the transition region in the QCD phase diagram.

In an interview following the event marking 30 years of heavy-ion physics at CERN, Hans J. Specht was asked, "Is there a future for heavy-ion physics at the SPS?" To this, he responded:

"The SPS is still the best machine for performing precision studies of the different transitions—deconfinement and chiral phase transitions—within the QCD phase

diagram. There is no chance of ever beating the factor of 1000 superiority at any collider, be it RHIC or the LHC, simply due to the luminosity difference which is obtained in principle with fixed-target machines. There is an experiment, NA61, which is performing energy and atomic number scans. A future follow-up of the NA60 experiment, NA60 +, is also under discussion, with another factor of 100 increase in data quality. This would be the ultimate experiment to clarify the remaining open issues, such as the chiral transition and the onset and order of these transitions. "

NA60+ A Jewel in the Future of SPS Heavy-Ion Physics

Hans J. Specht affectionately referred to NA60 + as a "little jewel", with its potential to achieve dilepton statistics 100 times larger than those of NA60. This achievement would allow, for the first time, direct evidence of chiral mixing by measuring the in-medium spectral properties of the a1 meson—the chiral partners of the ρ-meson. The submission of the Letter of Intent for NA60 + to the SPSC by Gianluca Usai et al. was a moment of great satisfaction for Hans J. Specht. He viewed it as a significant step forward in the continuation of heavy-ion research at SPS.

Hans J. Specht's unwavering belief in the SPS's future potential underscores his legacy as a visionary—someone who never stopped looking ahead.

Music, Physics, and Neurophysiology

This concerns a particular passion in my professional life, which had a unique beginning 30 years ago. In 1986, the 600th anniversary of Heidelberg University, Hans Günther Dosch and I offered a lecture for students from all faculties titled "Helmholtz and Beyond—Physics and Music." With both of us on stage at the same time, we divided our roles between experiments (conducted by me, Hans J. Specht) and interpretation or theory (presented by Hans G. Dosch). This collaboration created an unexpectedly lively atmosphere that actively engaged the audience (Fig. 3.57).

Fig. 3.57 Lecturing in 1986, on the occasion of 600th anniversary of the University of Heidelberg, 'Helmholtz und Danach: Physik und Musik.' From right to left: Hans Günter Dosch and Hans J. Specht. © FAIR-GSI. All rights reserved

Live Experiments and Musical Demonstrations

The experiments included Helmholtz's apparatus in Heidelberg, such as sirens and a complete set of spheres, as well as electronic generators for numerous psychoacoustic paradigms, analyzed live using rapid Fourier analysis and time-domain visualization with an oscilloscope. Along with a projector for Dosch's explanations, this setup required three screens to be used simultaneously. Live musical performances also played a significant role, with the piano played by me and the flute by Dosch for quick demonstrations of various kinds. A wide range of other instruments was also featured, with various colleagues from the audience eagerly stepping in to participate. Overall, it was an unforgettable event which quickly gained attention beyond Heidelberg.

Invitations soon followed, starting with a colloquium in Bielefeld in 1987, which also attracted listeners from the nearby Detmold University of Music. Over time, there were more than 20 invitations for lectures around the world, including prestigious venues at GSI, DESY, and CERN, as well as meetings of the Alexander von Humboldt Foundation, the Loeb Lectures at Harvard (6 h), and universities in Munich and Vienna, not to mention the Einstein Lectures in Berlin. There were also workshops organized by the Leopoldina, along with general lectures at the Deutsches Museum in Munich and in conjunction with music festivals such as Verbier in Switzerland and the Beethoven Festival in Bonn, where we lectured in the chamber music hall of the Beethoven House (Figs. 3.58, 3.59 and 3.60).

These events featured live performances, often culminating in short concerts, with movements from string quartets or other classical works performed by talented amateur musicians and students from local music conservatoires. Notably, in Berlin, Alban Gerhardt, one of the leading cellists in the world today, took part in our event.

The rapidly growing number of invitations was no coincidence: while the almost playful style was preserved, the learning processes for achieving a deeper understanding through original literature were profound. This was further enhanced by two weekly lecture series in 1994 and 2000, which included participants from the Mannheim University of Music and Performing Arts. During the preparation of psychoacoustic experiments, we realized that our own ears did not always perceive what the psychoacoustic literature described. This discrepancy became the strongest

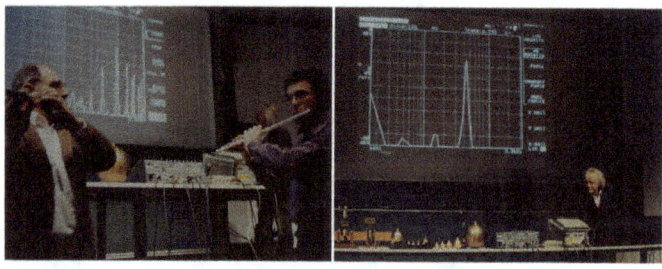

Fig. 3.58 Lecturing at GSI "On the Physical Cause of Harmony and Disharmony" (H. von Helmholtz) in 1994. © FAIR-GSI. All rights reserved

3   Hans Joachim Specht (in His Own Words) …

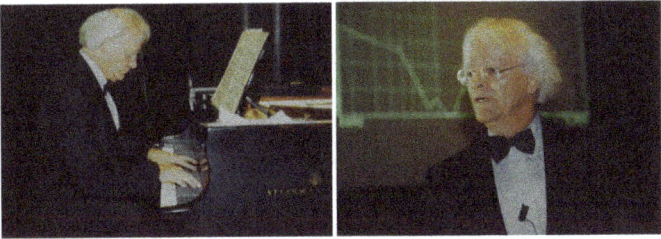

Fig. 3.59 Lecturing on "Musical Harmony—Physics, Physiology, Psychology" in October 2000 during the International Beethoven Festival in Bonn, held in the Chamber Music Hall of the Beethoven House. © Hans J. Specht. All rights reserved

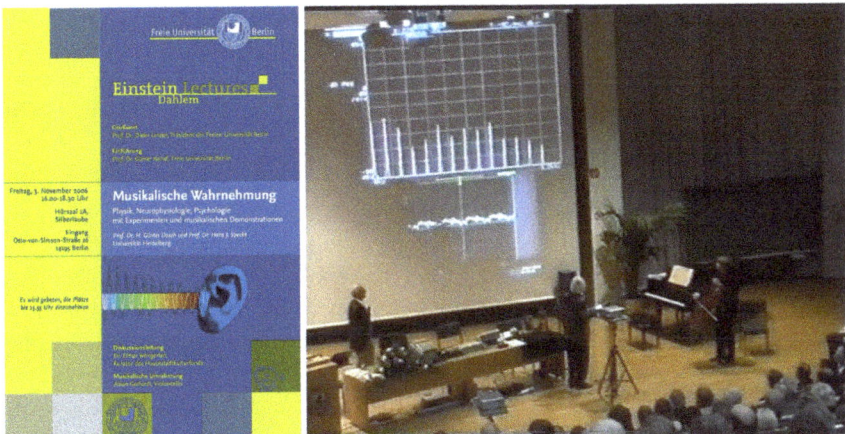

Fig. 3.60 Einstein Lectures Dahlem on 'Musical Perception—Physics, Neurophysiology, Psychology' in Berlin, 2006. Cello performance by professional musician Alban Gerhardt, one of the world's leading cellists. © Hans J. Specht. All rights reserved and © Bernd Wannenmacher, Freie Universität Berlin. All rights reserved

motivation for conducting our own research, seeking to understand the brain's role in musical perception.

Psychoacoustic tests had already been conducted in the lecture hall with students, such as the perception of consonance and dissonance in sinusoidal tones. Around the year 2000, we established a collaboration with Prof. M. Scherg's group "Section of Biomagnetism" at the Department of Neurology, University of Heidelberg, which specializes in magnetoencephalography (MEG), but also utilizes standard methods like magnet resonance tomography (MRI) in cooperation with the Department of Neuroradiology. MEG measures the electrical currents that occur in the brain during perception and cognitive processes by detecting the magnetic field gradients they produce. In contrast, MRI assesses local anatomy in terms of white and gray matter. This connection led to several years of intensive collaboration, later continued by

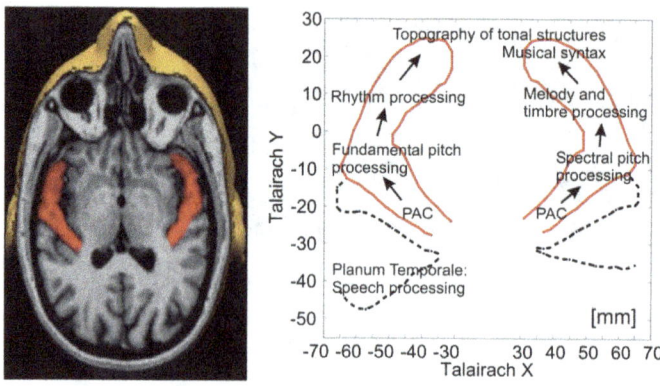

Fig. 3.61 Top view of the left and right auditory cortex of P. Schneider, highlighting Heschl's gyrus, which includes areas for sound and music processing (the first Heschl convolution is colored in red). PAC: primary auditory cortex; arrows illustrate the hierarchical organization, accessible through temporal analysis using magnetoencephalography. © Hans J. Specht. All rights reserved and © 2005 Springer Nature America, Inc. All rights reserved

A. Rupp as the head of the Section of Biomagnetism. The rigorous methods we brought from physics contributed significantly to this interdisciplinary effort.

At the center of these investigations from the very beginning was Heschl's gyrus, hosting the primary auditory cortex in the brain, where the early processing of musically relevant attributes like pitch and timbre occurs. The first publication stemming from this transdisciplinary research was the doctoral thesis by P. Schneider, a physicist and professional musician, who had previously completed his diploma thesis in psychoacoustics with us and continued to be mentored by our team of physicists thereafter (Fig. 3.61).

Surprising Discoveries in Auditory Processing

Our 2002 publication in Nature Neuroscience (Vol. 5, p. 688) revealed surprising and completely unexpected results. Using MRI and MEG, we compared the way non-musicians, professional musicians, and amateur musicians process harmonic complex tones in the auditory cortex. We discovered that both anatomical and neurophysiological data showed significantly increased attributes in professional musicians. Specifically, the gray matter volume of Heschl's gyrus was on average 130% larger in professional musicians than in non-musicians (Figs. 3.62 and 3.63).

Figure 3.64 (from Nature Neuroscience 5 (2002) 688) summarizes these surprising findings. For the early P30 response analyzed here, we found that the dipole amplitude of this primary MEG signal correlates both with the gray matter volume of the first transverse Heschl's gyrus including the primary auditory cortex, and the tonal score of the AMMA test developed by Edwin E. Gordon. This test is widely used

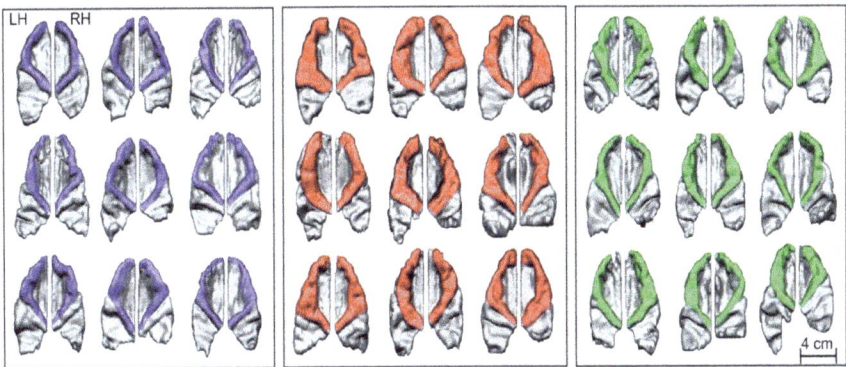

Fig. 3.62 Example of results. Size of the auditory cortex, reconstructed from high-resolution MRI scans for non-musicians (left), professional musicians (middle) and amateurs (right). The size of the first transverse Heschl's gyrus (colored) varies systematically between groups and hemispheres, with the largest observed in professional musicians. © Hans J. Specht. All rights reserved and © 2005 Springer Nature America, Inc. All rights reserved

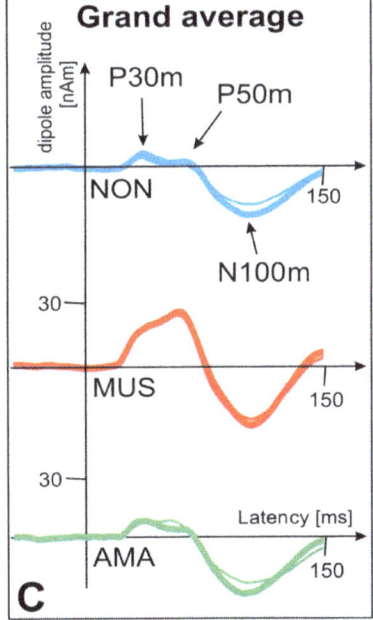

Fig. 3.63 MEG signals: group-averaged auditory evoked responses of the right (thicker curve) and left (thinner curve) auditory cortex in non-musicians (blue), professional (red), and amateur musicians (green) while listening to harmonic complex tones. Each curve illustrates the primary P30, secondary P50, and N100 response patterns occurring approximately 30, 50, and 100 ms after tone onset. The first 150 ms following tone onset are shown. © Akustik Journal, (2018). All rights reserved and © Peter Schneider. All rights reserved

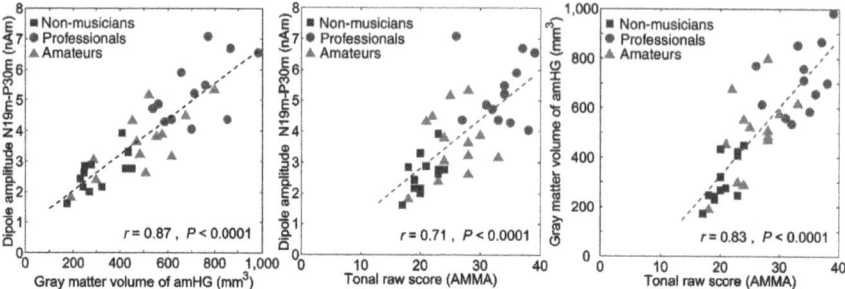

Fig. 3.64 Correlation between neurological measures and the musical aptitude measured with the AMMA test by Edwin E. Gordon. © 2002 Springer Nature America, Inc. All rights reserved

today in conservatories around the world. Professional musicians were shown to have a Heschl's gyrus in the P30 region which is twice as large as that of non-musicians, while amateur musicians, depending on their aptitude, span a broad range of values between the two. Such significant effects linking subjective attributes (such as musicality) with objective brain measurements have been little documented so far. Interestingly, this publication in Nature Neuroscience is my most cited paper with 1006 citations to date, together with the 1995 CERES result.

The referees of this study engaged in intense debates with us regarding the underlying cause of the observed correlation: was it innate talent or the result of extensive practice? In fact, no correlation was found between the P30 neural component and the number of hours practiced. However, in a subsequent measurement, we identified a stronger correlation between practice time and the P50 component. This measurement showed a monotonic increase in MEG signals with practice time, reaching up to five times larger in professional musicians who had practiced for 5–10 h a day over the past decade. In contrast, amateurs like myself, who practiced between 0 and 2 h a day, exhibited only small signals (Fig. 3.63).

The Dual Perceptions of Pitch

Another publication, in Nature Neuroscience 8 (2005) 1241, addresses the intriguing topic of pitch recognition. Since the 1930s, it has been known that pitch is not determined solely by the fundamental tone of a harmonic series, but rather by its mere existence. This principle underlies the perception of bass tones in various musical instruments, such as the piano or in artificially truncated sound reproduction. Through psychoacoustic tests (with incomplete sound complexes, such as those with only the harmonics n = 3, 4, 5), we observed two distinct groups of listeners: "fundamental-tone listeners," who perceive the missing fundamental tone even if not physically present, and "overtone or special pitch listeners," who perceive higher-pitched harmonics. The frequency distribution of these two types does not have a peak in the middle; instead, individuals predominantly identify as one or the other.

Our neurophysiological investigations uncovered a fascinating hemispheric specialization. Fundamental-tone listeners exhibited a larger left Heschl's gyrus, while overtone or spectral-pitch listeners had a larger right Heschl's gyrus. This finding aligned with previous research showing that pitch recognition is predominantly associated with the left hemisphere (see the reference list in *Nature Neuroscience 8 (2005) 1241*). The musical implications of the unique role of harmonic series and their processing extend from Rameau's concept of "basse fondamentale" to, potentially, the choice of instrument made by professional musicians, which may be influenced by their listener type.

Over the past 30 years, this journey through music, physics, and neurophysiology has revealed fascinating connections between the physical properties of sound and the complex workings of the brain. From Helmholtz's pioneering insights to our own discoveries in neurophysiology, we have deepened our understanding of how the brain perceives music—bridging the gap between sound vibrations, psychological experience, and the neurological processes which give rise to harmony, dissonance, and pitch recognition. As we continue to explore these connections, I look forward to uncovering even more about the profound impact of music on human experience (Fig. 3.65).

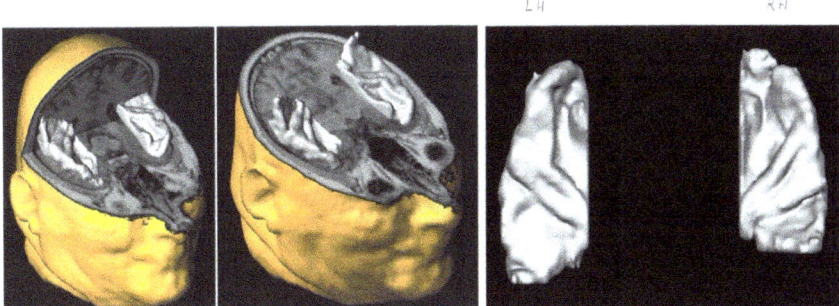

Fig. 3.65 Hans J. Specht—Physicist and "Pianist": 3D reconstruction (left and middle panels) and top view (right panel) of Hans J. Specht's right and left auditory cortex, including his distinct and exceptionally large bilaterally duplicated Heschl's gyri, which are responsible for sound and music processing, as typically found in professional pianists. The anterior and posterior Heschl's gyrus are separated from each other by a clearly visible 'sulcus intermedius.' The 3D reconstruction was created by half-automatic segmentation techniques from a structural isovoxel 1 mm MR image at the Heidelberg Head Clinic. © Hans J. Specht, All rights reserved

Selected Publications—A Guide

Neuroscience of musical relevance

- Morphology of Heschl's gyrus reflects enhanced activation in the auditory cortex of musicians

  P. Schneider, M. Scherg, H.G. Dosch, H.J. Specht, A. Gutschalk, and A. Rupp

  Nature Neuroscience 5 (2002) 688–694

  Musical aptitude ("talent") is visible in the size of the relevant gray matter region in Heschl's gyrus, with a dynamic range of a factor of 2 (> 1000 citations). Extended part of Ph.D. thesis by P. Schneider (supervisors Dosch/Specht).

- Structural and functional asymmetry of lateral Heschl's gyrus reflects pitch perception preference

  P. Schneider, V. Sluming, N. Roberts, M. Scherg, R. Goebel, H.J. Specht, H.G. Dosch, S. Bleek, C. Stippig, and Andre Rupp

  Nature Neuroscience 8 (2005) 1241–1247

  The left/right structural and functional asymmetry of Heschl's gyrus correlates with the relative dominance of fundamental pitch versus overtone/spectral-pitch perception.

  The specific psychoacoustic test ("pitch perception preference test," Schneider et al., 2005, developed according to the earlier concept of Schmorenburg (1971)) used in this paper was later offered with a test CD by the German magazine AUDIO (January 2006, 8–16) to a readership of about 100 000. About 5700 sent back the test results for analysis by the HD group (P. Schneider).

  Correlations between listener types and preference for specific hardware, the driving motivation for the mass test, were not found. The global result agrees with the one in Nature and is essentially independent of age, gender, and musical experience.

Lectures on Music, Physics, and the Brain around the World

H.G. Dosch and H.J. Specht

1987 **"Helmholtz und danach—Physik und Musik"**
Physikalisches Kolloquium, Fakultät für Physik, Universität Bielefeld

1994 **Über die physikalische Ursache der Harmonie und Disharmonie (H. von Helmholtz)**
Physikalisches Kolloquium, Fakultät für Physik und Astronomie, Uni Heidelberg, zus. mit Helmholtz-Symposium 1994 (100. Todestag von H.v.H), 11 Feb 1994

1994 **Über die physikalische Ursache der Harmonie und Disharmonie (H. von Helmholtz)**
GSI Kolloquium, GSI Darmstadt (Fig. 3.66)

1997 **Musical Harmony & Physics**
Jahrestagung der Alexander von Humboldt-Stiftung, Bamberg

1998 **Psychophysics of Musical Consonance**
CERN Lectures on Science and Society, CERN, Geneva, 22 January 1998

1998 **DESY Kolloquium, DESY Hamburg**

1998 **Physikalisches Kolloquium, Fakultät für Physik, TU Karlsruhe**

1999 **Musical Harmony—Physics, Physiology, Psychology, Colloquium**
Musical Pitch: Temporal versus Spectral Perception, Lecture I
Physics of Musical Instruments and Pitch Perception, Lecture II
Harvard University, Cambridge, Morris Loeb Lectures in Physics, 12–15 April

1999 **Musical Harmony & Physics**
Verbier Music Festival & Academy, 24 July 1999 (Fig. 3.67)

Fig. 3.66 Experiments during the *Lecture on Music, Physics, and the Brain* at GSI, featuring Helmholtz's apparatus (sirens, spheres, and electronic generators) and real-time analysis of psychoacoustic paradigms using Fourier transforms and oscilloscope visualization. © FAIR-GSI. All rights reserved

Harvard University and Verbier Music Festival

Fig. 3.67 Favorite poster designs for many *Lectures on Music, Physics, and the Brain*, with examples from the Harvard University and Verbier Music Festival lectures. © Hans J. Specht, All rights reserved

1999 **Musikalische Harmonie—Physik, Physiologie, Psychologie**
GSI Kolloquium, GSI Darmstadt

2000 **Musikalische Harmonie—Physik, Physiologie, Psychologie**
Kolloquium der Münchner Physiker (TUM und LMU), Physik-Depart. TU, 7 Feb

2000 **Musikalische Harmonie—Physik, Physiologie, Psychologie**
Physik Event der ÖPG, Graz, September 2000

2000 **Musikalische Harmonie—Physik, Physiologie, Psychologie**
Begleitpr. Int. Beethovenfest Bonn, Kammermusiksaal Beethovenhaus, 5 October

2001 **Musikalische Harmonie—Physik, Physiologie, Psychologie**
Wiener Physikalisches Kolloquium, TU Wien, 12 November 2001

2002 **Musikalische Harmonie—Physik, Physiologie, Psychologie**
Heidelberger Akademie der Wissenschaften, Heidelberg

2004 **Currents in the Brain and Musical Perception**
International Leopoldina Symposium "Science and Music—The Impact of Music"
Deutsche Akademie der Naturforscher Leopodina, Halle, 13–15 May 2004

2004 **Musikalische Wahrnehmung und Ströme im Gehirn**
Physikalisches Festkolloquium, Fakultät fur Physik und Astronomie, Unversität
Heidelberg, aus Anlass des 80. Geburtstag von B. Stech, 17 December 2004

2005 **Musikalische Wahrnehmung und Ströme im Gehirn**
Science for Everybody, Deutsches Museum München, March 2005

2006 **Musikalische Wahrnehmung—Physik, Neurophysiologie, Psychologie**
Einstein Lectures Dahlem, Freie Universität Berlin, 3 November 2006

2007 **Musikalische Wahrnehmung—Physik, Neurophysiologie, Psychologie**
Fachbereich Physik, "400 Jahre Universität Giessen," Giessen, 7 May 2007

2013 **Töne, Klänge, Musikalische Harmonie**
Internationale Akademie Traunkirchen, Kolloquium und Workshop mit Seminarvorträgen von Studenten und Dosch/Specht; Organiser A. Zeilinger; 8 September 2013 (Fig. 3.68).

Fig. 3.68 Poster of the *Sounds, Tones, Musical Harmony* Colloquium at the International Academy Traunkirchen. © Internationale Akademie Traunkirchen. All rights reserved

Scientific Managing Director of GSI Darmstadt, 1992–1999

Pioneering Ion-Based Cancer Therapy and the Shaping of GSI's Long-Term Future

As my experience in leading experimental research groups grew, I became increasingly involved in numerous national and international advisory committees, often serving as chair. These roles gradually introduced me to the world of science management. The following are among my most cherished memories: 15 years on the Alexander von Humboldt Foundation's award committee; being part of the inaugural Board of Directors of the European Center for Theoretical Studies in Nuclear Physics and Related Areas (ECT*) in Trento, under the leadership of Ben Mottelson (Nobel Prize laureate); chairing CERN's Proton Synchrotron and SynchroCyclotron Committee (PSCC); and serving 10 years on CERN's Science Policy Committee (SPC), with the unique tradition of annual re-invitation since 2000. However, my most enjoyable committee experience was with the "LHC Safety Study Group" at CERN, tasked to evaluate the potential production of black holes and other fantastical scenarios at the LHC.

More relevant to my work at GSI were my contributions to four review panels and two ad hoc committees for the German Federal Ministry of Research and Technology (BMFT).

Taking on the Role of Scientific Managing Director (1992)

In 1992, I was fully thrust into a new role. I accepted, albeit with some hesitation, the position of Scientific Managing Director of GSI ("Wissenschaftlicher Geschäftsführer der GSI Darmstadt"). One of my first initiatives was to improve management efficiency; I advocated for a change from the rigid "Geschäftsführers/GmbH" structure of sole executive leadership. Instead, I established a directorial board similar to those at CERN and DESY, comprising directors for research (Volker Metag, from Giessen), accelerators (Norbert Angert, from GSI), infrastructure (Wolfgang von Rüden, from CERN), and administration (Hermann Zeitträger, GSI's Administrative Director) (Figs. 3.69 and 3.70).

The efficiency of communication was further improved by regularly involving leading scientists like Peter Armbruster, Rudolf Bock, and Jürgen Kluge in routine meetings. This collaborative structure allowed me, as chairman, to delegate many routine tasks, enabling smoother operation. Another key initiative I took right away was to elevate the status of the Scientific Council by adopting English as its official language, fostering a consistently international composition. This new structure significantly improved the efficiency of GSI's management (Figs. 3.71, 3.72 and 3.73).

Fig. 3.69 GSI experimental facilities in the 1990s. © FAIR-GSI. All rights reserved

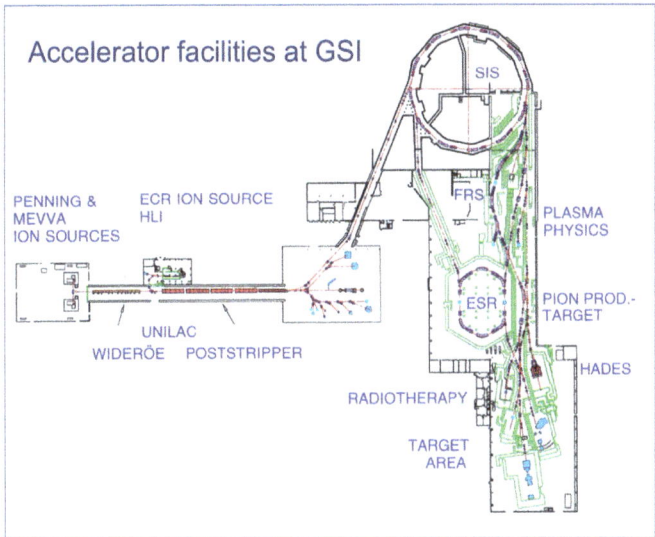

Fig. 3.70 GSI experimental facilities in the 1990s. The accelerator complex consisted of the linear accelerator UNILAC, the medium energy synchrotron SIS18, and the storage cooler ring ESR, together with several large spectrometers and advanced detector systems for basic research in nuclear and atomic physics, as well as applied studies in plasma physics, material research, biophysics, and radiotherapy. © FAIR-GSI. All rights reserved

Fig. 3.71 The four Scientific Managing Directors of GSI since 1969: Christoph Schmelzer, Gisbert zu Putlitz, Paul Kienle, and Hans J. Specht (from left to right). © FAIR-GSI. All rights reserved

Fig. 3.72 History of the development of GSI up to the 1990s with two milestone studies highlighted in red. © Hans J. Specht, All rights reserved

History in the Development of GSI up to 1990s

1959 Start of development of the UNILAC by C. Schmelzer et al., Heidelberg

1969 Foundation of GSI, Darmstadt

1976 First Uranium beams, initially up to 9 MeV/u, later increased to 20 MeV/u

1985 Approval of the SIS18/ESR Project

1990 Start of operation of SIS18/ESR, protons max. 4.5 GeV, U 1.0 GeV/u

1998 Start of studies towards further expansion of the GSI facilities

International Growth and New Projects

During the 1990s, GSI operated with approximately 700 employees, including 300 scientists and engineers. The facility served around 1000 external users, 400 of them being international users from over 150 institutes across 25 countries, highlighting **the growing international use of GSI's facilities**. With an annual budget of 125 million Deutsche Marks—100 million allocated for operations and 25 million for investments—GSI's primary objective was to fully exploit the extensive opportunities offered by the UNILAC accelerator and the newly established SIS18 and ESR facilities.

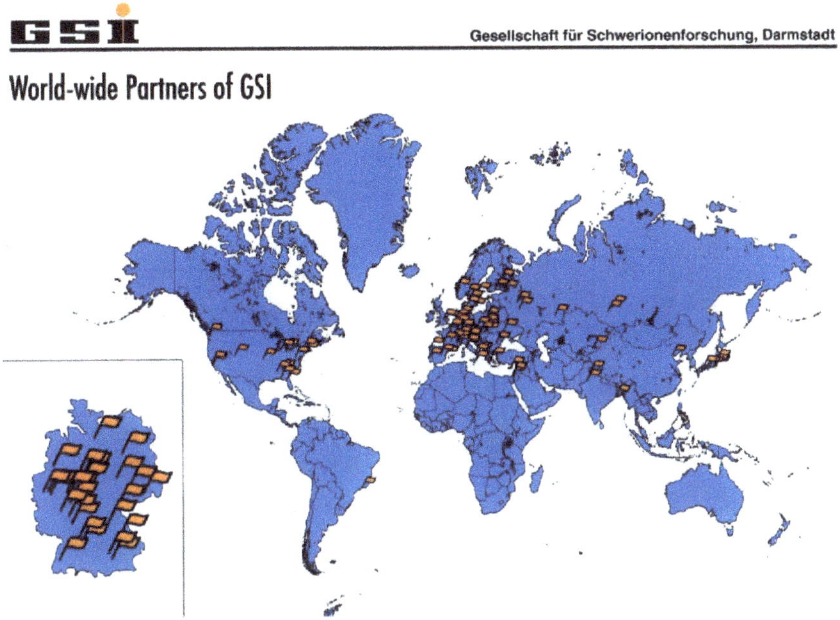

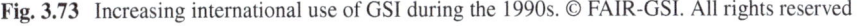

Fig. 3.73 Increasing international use of GSI during the 1990s. © FAIR-GSI. All rights reserved

The Balance of Joy and Frustration

The 7 years, I spent commuting between Heidelberg and Darmstadt—balancing management responsibilities, scientific oversight in Darmstadt, and my personal research in Heidelberg/Geneva—lasted until the fall of 1999. Of course, the latter activities occurred outside my regular, already packed schedule. In retrospect, as I noted in my official farewell speech, I often described these 7 years as like walking a tightrope between joy and frustration.

Among the sources of joy during this period was the significant progress made in fundamental research, GSI's core mission, across all research fields. This is exemplified by the discovery of over 500 new and far from stable isotopes. The crowning achievements were the discoveries of superheavy elements with atomic numbers 110, 111, and 112—darmstadtium (Ds-110), roentgenium (Rg-111), and copernicium (Cn-112)—between 1994 and 1996, following the discoveries from the initial series of experiments: bohrium (Bh-107) in 1981, meitnerium (Mt-109) in 1982, and hassium (Hs-108) in 1984. In the second series of experiments, between 1994 and 1996, during my time as GSI director, I had a significant influence in securing sufficient beamtime for these crucial experiments.

These discoveries were celebrated with visits from high-ranking officials and ministers from Bonn, among other events, bringing international recognition to GSI and enhancing its reputation in the scientific community (Figs. 3.74 and 3.75).

Fig. 3.74 A unique occasion at GSI in 1996, with the three 'fathers' of all the elements above U-92 (the 'transuranium' elements): Peter Armbruster, Glenn T. Seaborg (Nobel Prize laureate), and Yuri Oganessian with Hans J. Specht. © FAIR-GSI. All rights reserved

Fig. 3.75 The element with atomic number 110, discovered in 1994, was officially named darmstadtium at a special ceremony at GSI in 2003. Left: Federal Minister for Education and Research, Edelgard Bulmahn, alongside Peter Benz, Mayor of Darmstadt (from left to right). Right: GSI Scientific Directors Gisbert zu Putlitz, Paul Kienle, and Hans J. Specht (from left to right). © FAIR-GSI, with a photo credit to A. Zschau. All rights reserved

The Milestone Year 1994: Celebrating GSI's 25th Anniversary

The year of 1994 was exceptional for GSI, a fitting way to celebrate the 25th anniversary of the laboratory. A number of important results were obtained, capturing not only the interest of experts but also widespread public attention. Among these were the detection of the doubly magic nucleus ^{100}Sn and the discovery of two new elements, $Z = 110$ and $Z = 111$, later named darmstadtium (Ds-110) and roentgenium (Rg-111). These discoveries were made possible by the upgraded SHIP spectrometer at the UNILAC. Together with other breakthroughs in nuclear physics and the announcement of an upcoming therapy unit for cancer treatment using ions, these achievements served as a fitting tribute to this milestone year, further reinforcing GSI's role in advancing nuclear physics and demonstrating its commitment to both pioneering basic research and practical applications (Fig. 3.76).

After I had welcomed an audience of more than 500 participants, Hartmut Grübel, Chairman of the GSI Board from the BMBF, conveyed warm greetings and best wishes from Jürgen Rüttgers, the Federal Minister for Education, Science, Research, and Technology. Other notable speakers included Christine Hohmann-Dennhardt, Hessian Minister of Science and Art, Joachim Treusch, AGF Chairman and Director of the Jülich Forschungzentrum, Peter Benz, Mayor of Darmstadt, Chris Llewellyn Smith, CERN Director-General, and Wolfgang Frühwald, President of the "Deutsche Forschungs Gemeinschaft" (DFG).

Dirk Schwalm, Managing Director of MPIK Heidelberg, summarized the scientific highlights of GSI's first 25 years, including its unique accelerator facilities (UNILAC, SIS18, ESR), the discovery of five superheavy elements, and the promising future of tumor therapy with ion beams. He stated that many dreams from

Fig. 3.76 Twenty-fifth anniversary of GSI—12 May 1995. Hans J. Specht, Scientific Managing Director of GSI, welcoming an audience of more than 500 to celebrate the laboratory's 25th anniversary. © FAIR-GSI, with a photo credit to A. Zschau. All rights reserved

the founding period of GSI had become a reality, and even what one had not dared to dream of was on its way to becoming reality, such as tumor therapy with ion beams. In a very well received talk, Wolfgang Frühwald, discussed the relationship between fundamental and applied research. He made the point that the rules for technological innovation cannot be applied to fundamental research, which in contrast must continue as an open science. But, while innovation and fundamental research remain separate, each nevertheless needs the other to move forward.

Pioneering Applied Research at GSI

While GSI was well known for its achievements in fundamental research, I became increasingly interested in its potential for applied research, since certain specific conditions at GSI made it unique thanks to GSI's cutting-edge facilities and expertise in ion-beam technology, offering capabilities unmatched elsewhere. Thus, these projects received the highest priority, although unfortunately not always with the approval of fundamental purists.

During this period, two major application-oriented projects were launched:

- The pioneering use of carbon ions for tumor therapy, treating patients on-site using the SIS18 accelerator.
- Development of the Petawatt High-Energy Laser for Heavy-Ion Experiments (PHELIX), designed to explore various fields related to plasma physics and complementing heavy-ion research.

Both initiatives demonstrated GSI's dedication to transferring cutting-edge technologies into impactful real-world applications.

Turning Basic Science into Service for Humanity

Cancer Therapy with Ion Beams

One of the most transformative initiatives—and a dream I pursued throughout my tenure as Scientific Managing Director—was the development of innovative radiation therapy using accelerated carbon ions. This breakthrough made it possible, for the first time, to treat specific, well-localized tumors, especially those in the brain, which were neither operable nor accessible to conventional radiation therapy. This approach offered patients a real chance for long-term survival, making it a life-saving innovation.

When news spread of my appointment at GSI, Ugo Amaldi, a strong advocate for hadron cancer therapy (including the PIMMS design at CERN in the early 1990s), remarked that I could not possibly turn down such an offer. But it was one of my colleagues from Heidelberg, J. Heintze, who contributed the best anecdote:

"Mr. Specht, as an enthusiastic researcher, you are only wasting your time getting involved in years of management, but if you succeed in bringing a cancer therapy machine to Heidelberg, it will have been worth it."

Convinced of the immense potential and technical feasibility of ion-beam therapy in radiation medicine, I made it a top priority within GSI's scientific and technical program. The medium-term goal was to establish experimental tumor therapy for patients on-site at GSI. This decision, which was hotly debated within the scientific community at the time, including the previous management, was a courageous step which required strategic planning and interdisciplinary collaboration.

The Beauty of the Bragg Peak

In contrast to X-rays and electrons, heavier charged particles exhibit minimal lateral and angular scattering. These particles travel in virtually straight lines and stop at a specific penetration depth, delivering a high dose precisely at the end of the ion path, a phenomenon known as the Bragg peak (Fig. 3.77).

The Bragg peak allows for concentrated energy deposition at the tumor volume while minimizing damage to surrounding healthy tissues. The optimum dose of heavy ions can be precisely localized within the tumor, ensuring effective treatment, while protecting healthy cells. This approach can reach tumors wherever they are located in the body.

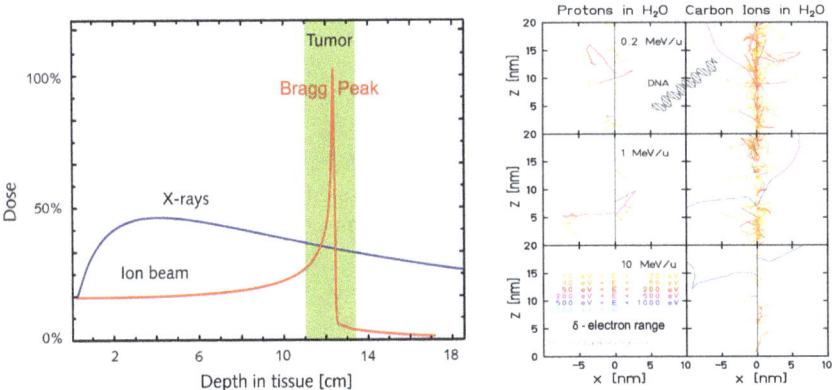

Fig. 3.77 Left: Depth dose distribution illustrating the unique characteristics of the Bragg peak. The inverted dose profile for carbon-12 ions (^{12}C) is compared to X-rays, showing how carbon ions deliver concentrated energy within the target area, while minimizing the exit dose, a hallmark of hadron therapy. Right: Dose distribution at the nanometer scale, illustrating the effects of DNA damage. Single-strand breaks caused by protons and X-rays are generally reparable, whereas multiple strand breaks induced by ^{12}C ions are more complex and typically irreparable, effectively damaging radiation-resistant cancer cells. © FAIR-GSI. All rights reserved

Key Collaboration and Government Support

The most crucial decision was to establish a groundbreaking pilot project involving patient irradiation directly on the GSI premises. Realizing such a patient irradiation facility at GSI's accelerator complex required the formation of a powerful research alliance, bringing together institutions with scientific and technical expertise but also, crucially, with the necessary medical proficiency. Within just 8 months of my arrival, we had assembled a comprehensive 100-page proposal, which became a model of interdisciplinary collaboration between GSI, the Radiological Clinic of the University of Heidelberg, and the German Cancer Research Center (DKFZ) in Heidelberg. The project included state-of-the-art irradiation and control rooms designed to closely resemble what would later become the Heidelberg Ion-Beam Therapy Center (HIT). This interdisciplinary collaboration was not only vital but also transformative (Fig. 3.78).

As Professor Gerhard Kraft recalls: *"GSI did not want to be involved in the hadron therapy project except for beam production. The great fear was that ion therapy would be too successful and could overshadow the nuclear and atomic physics program. The view changed completely in the spring of 1993 when Hans J. Specht took over the directorship of GSI. He wanted to see the first patient treated within the first four years of his directorship. He was willing to support the project with the highest priority in all technical, physical, and radiobiological details, in whatever areas GSI had the necessary experience. Independently of any new proposal for the [federal and state] governments, the construction of a medical radiation facility began immediately in May 1993."*

In retrospect, the GSI Supervisory Board, especially its chairman Hartmut Grübel, and later Hermann Schunck, BMFT, deserve the highest praise for their willingness, not only to approve the project but also to make it almost their own, providing continued support.

In May 1993, we submitted a formal funding request to the Federal Minister for Research and Technology, Dr. Paul Krüger, seeking financial support and additional positions to bring the pilot project to life. The letter was co-signed by myself on behalf of GSI, Professor Dr. Dr. Michael Wannenmacher, Medical Director of Heidelberg University Hospital, representing the Radiological Clinic of the University of Heidelberg, and Professor Dr. Harald zur Hausen, Scientific Director of the German Cancer Research Center (DKFZ) and later recipient of the 2008 Nobel Prize in Medicine, representing the DKFZ in Heidelberg. Between 1993 and 1997, numerous visits by German Federal and State Ministers were largely motivated by their interest in this pioneering pilot project on cancer therapy. I owe the next wonderful quote to Professor zur Hausen: "Mr. Specht, we are already on board, for your sake and your colleagues' sake, but without much belief in the long-term perspectives of such methods: in 10 years, we will be at a point where we don't need such methods anymore!" (Fig. 3.79).

I often attended routine technical meetings with the main project participants, reinforcing the project's priority status within GSI. The research community respected this commitment, especially as my decision to favor parallel beams for both therapy

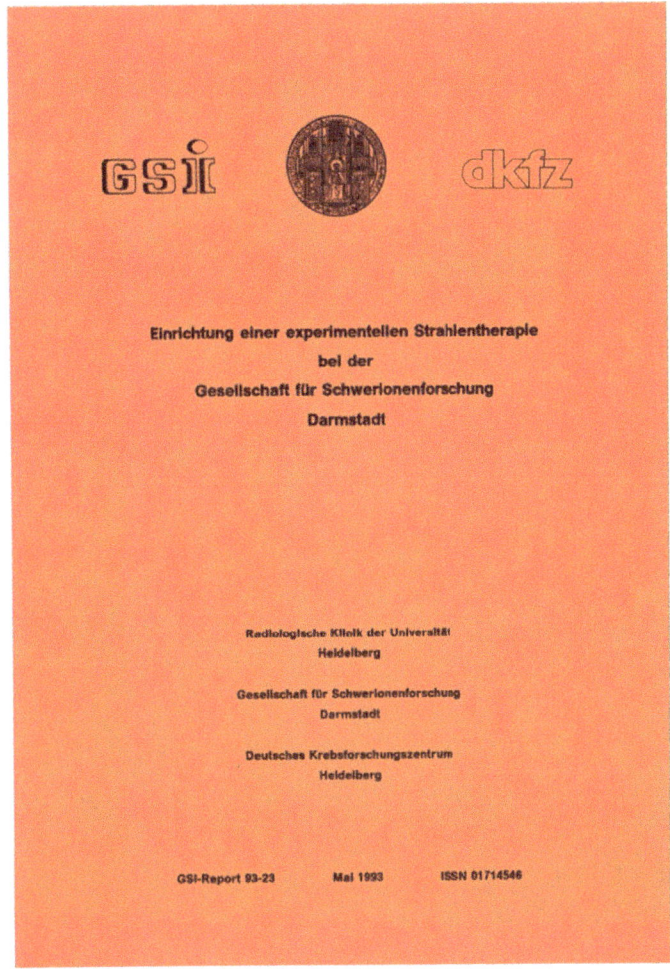

Fig. 3.78 Proposal for the pilot project with 12C ions at GSI, May 1993. © FAIR-GSI. All rights reserved

and research from the start helped avoid one of the fundamental mistakes made at Berkeley. I also fondly remember the warm and constructive collaboration with Michael Wannenmacher, and Jürgen Debus, the local medical project leader and M. Wannenmacher's successor. Harald zur Hausen also provided steadfast support for the project, despite his very different scientific focus (Fig. 3.80).

Technological Advancements and Milestones (1993–1997)

From 1993 to 1997, during the 4-year preparation phase, several technological innovations were developed, including the three-dimensional intensity-controlled raster scan technique, which enabled irradiation with precisions down to the millimeter.

Fig. 3.79 Dr. Paul Krüger, the Federal Minister for Research and Technology, visited GSI in April 1994. Right panel, from left to right: H. J. Specht, Minister Krüger, V. Metag, G. Kraft, and M. Wannenmacher. © FAIR-GSI, with a photo credit to A. Zschau. All rights reserved

Fig. 3.80 Visit of the Federal Research Minister Jürgen Rüttgers in 1995. From right to left: Federal Research Minister Jürgen Rüttgers, V. Metag, Prof. Dr. Evelies Mayer, Hessian Minister for Science and Arts, Professor Gerhard Kraft, Hans J. Specht, Andreas Storm, German Member of Parliament. © FAIR-GSI, with a photo credit to A. Zschau. All rights reserved

This technique allowed critical risk organs, such as the optic nerves or brainstem, to be entirely spared, even when located near tumors (Fig. 3.81).

These developments, driven by the research of Gerhard Kraft and Thomas Haberer, were complemented by a multi-year research program on cell behavior under radiation damage. Furthermore, the operational safety and stability of the GSI accelerator facilities were significantly improved, forming an indispensable prerequisite for medical applications. The GSI Biophysics Group succeeded in developing biologically optimized irradiation planning for the therapeutic use of ion beams (Voxelplan + LEM). Additionally, online quality control, including *in-situ* PET-diagnostics used

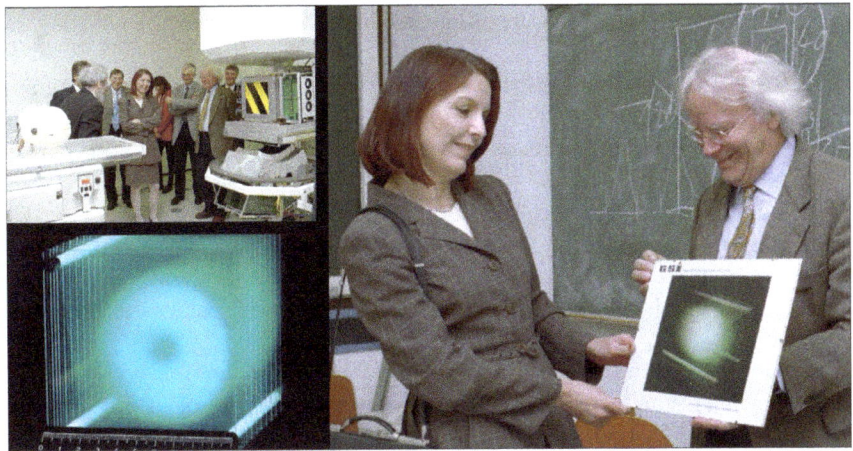

Fig. 3.81 C. Hohmann-Dennhardt, Hessian Minister for Science and Art, visited GSI in 1997. She received a small token illustrating the GSI beam-scanning technique. This uses raster scan, a method which allows tumors of any shape to be irradiated with millimeter-level accuracy. In this demonstration, plastic sheets immersed in water were irradiated in a doughnut shape. The biological effect along the path of the incoming ions is minimal, as the ions deposit the bulk of their energy in the target volume, ensuring maximal therapeutic effect while sparing surrounding tissue. © FAIR-GSI, with a photo credit to A. Zschau. All rights reserved

for the first time, provided millimeter-level precision, another significant breakthrough which enhanced patient outcomes. The performance of the facility was improved in terms of flexibility and stability, surpassing previous capabilities. As a result, the GSI experimental program could continue in parallel with the experimental therapy project without destructive interference, addressing a major concern for many experiments stemming from experience at Berkeley.

The irradiation cave and local control room were set up under the guidance of the Heidelberg Clinic to ensure a safe and "homely" environment. From the outset, procedures were designed to be familiar to the doctors and medical technical assistance staff from Heidelberg, fostering a sense of comfort and familiarity. This thoughtful design had an additional advantage: every visitor—whether a professional or a government minister, many of whom visited annually—felt as if they were entering a clinical facility rather than the experimental hall of a large physics institute. This did much to ensure that the clinic's competence in this massive endeavor was generally accepted. The technical preparations were managed by the GSI infrastructure team, which I successfully unified after more than 15 years of fragmented responsibilities among leading scientists.

Treating the First Patients with Carbon Ions

Achieving the realization of these novel technologies, building the pilot facility, and conducting intensive functional and safety tests within just 4 years was an immense challenge. This concentrated effort culminated in the historic moment of treating the first patient at GSI (Fig. 3.82).

When the first treatment was successfully completed, the space around the therapy cave filled with all the people who had worked so hard to make this treatment possible, and we celebrated this first success. This kind of therapy continued at GSI for over 10 years.

During the treatment process, the patients either made a daily bus trip or stayed in nearby hotels. One particularly memorable patient rode his bicycle several hundred kilometers from Switzerland, stayed at GSI during the treatment, and, after 20 fractions, cycled back home. This example illustrates how well patients tolerated the treatments.

The German government played a crucial role in supporting this breakthrough. Initially, health insurance providers were hesitant to take the risk of covering the first patients treated at GSI. In response, the government made the bold decision to provide insurance coverage for these patients for 30 years, demonstrating its strong commitment to the success of this novel technology. This government support ultimately paved the way for broader acceptance and insurance coverage of ion therapy. The successful treatment and survival of these early patients validated the government's decisive action, further consolidating the impact of ion-beam therapy.

This spectacular success also opened the way to the development of the clinical ion therapy machine in Heidelberg (Figs. 3.83 and 3.84).

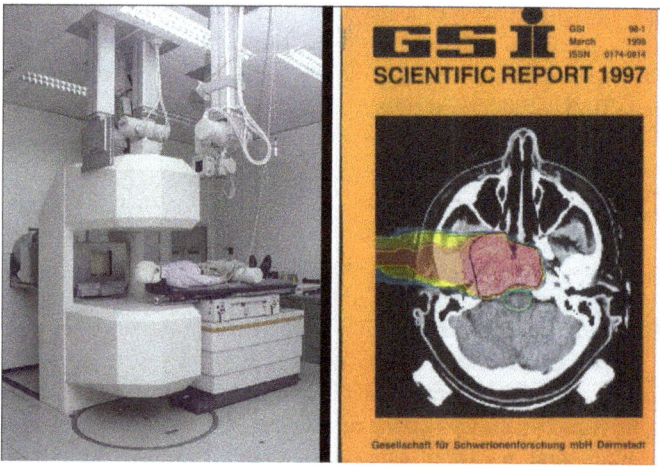

Fig. 3.82 Left: The GSI carbon-ion therapy facility treats its first patient on Saturday, 13 December 1997. The first patient suffered from a cranio-cervical chordoma which had been diagnosed in 1995. The patient underwent a surgical resection that same year, and suffered from tumor regrowth in 1997. A PET camera was positioned above and below the head during the treatment. Right: Physical dose resulting from the biological optimization for a single irradiation field used for the first patient. The isodose lines correspond to 20, 40, 60, 80, 90, and 100% of the maximum dose. © FAIR-GSI. All rights reserved

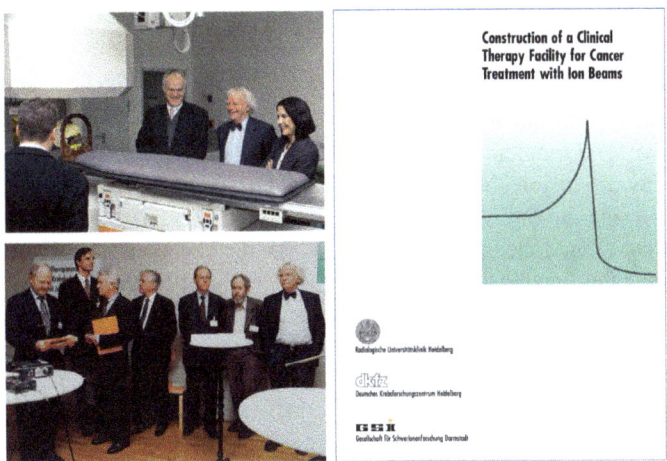

Fig. 3.83 Inauguration of the cancer therapy project at GSI on 15 September 1998. Top left from left to right: Dr. T. Haberer, Technical Director of HIT, J. Rüttgers, Federal Research Minister, H.J. Specht, and C. Hohmann-Dennhardt, Hessian Minister for Science and Art. Bottom left: M. Wannenmacher, J. Debus, M. Siebke, H. zur Hausen, U. Amaldi, G. Kraft, and H.J. Specht. *Right* Official project proposal for HIT Heidelberg, handed by H.J. Specht to Minister Rüttgers during the inauguration event. © FAIR-GSI. All rights reserved

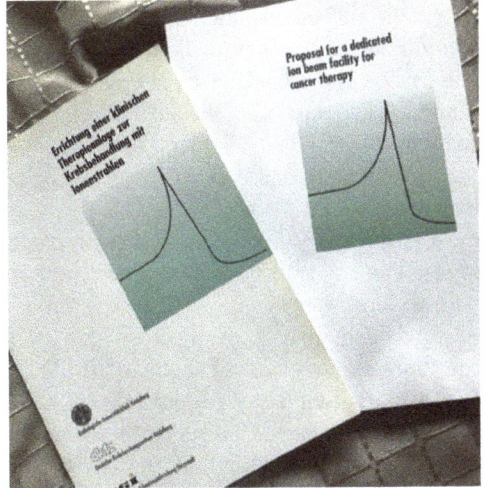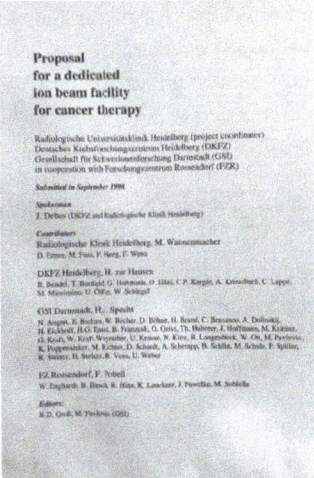

Fig. 3.84 Proposal and brochure for the first European Cancer Therapy Clinic using ion beams—HIT in Heidelberg. © FAIR-GSI. All rights reserved

Personal Reflection

The first patient irradiation took place in December 1997, and I still regard that day—from the tense atmosphere to the final announcement of success—as undoubtedly the most profound moment of my professional life. The actual irradiation occurred on a Saturday, a decision I had insisted upon after a very bitter dispute with the works council, ensuring the day would be entirely free of spectators. Only those directly involved were allowed access to the cave and the medical control room, not even I.

I wandered through the facility, speaking only to the operators in the main control room, who were monitoring the SIS18 and the beam delivery to the cave. Here, one could clearly sense the nervousness, evident in the chain-smoking and awkward jokes. Then, at last, the reassuring signal came: "Successfully completed, no incidents." As I made my way to the cave, the team met me halfway and we embraced one another, our faces filled with smiles and joy. There was hardly a happier hour in my whole professional life than after that first patient irradiation.

In December of the same year, two patients with irradiation-resistant tumors in the base of the skull region were treated by four and five fractions of carbon ions, respectively. These were the first patient irradiations ever made in Europe with heavy ions. Two technical novelties were utilized for the first time in therapy: the three-dimensional raster scan for direct control of the beam dose, and on-line positron emission tomography (PET) for direct control of the beam position. The patient treatments were performed to the utmost perfection and demonstrated the high reliability of the GSI accelerators as well as the millimeter precision of the new technique. The successful start of patient irradiations was a prerequisite for a clinical study which began in the summer of 1998, aimed at demonstrating the advantages of the new therapy for selected medical indications. Over the next 5 years, some hundred patients with irradiation-resistant tumors in the brain and base of the skull region were treated at GSI as part of this clinical trial.

In parallel with the clinical studies, a proposal for a new clinic was prepared, aiming to develop a dedicated therapy unit in order to transfer the new method and technology as soon as possible to a German hospital. The therapy project thereby demonstrated once again how fundamental research, together with technical innovations, could eventually lead to important practical applications. At the same time, it represented the very fruitful interdisciplinary cooperation with our project partners, the Radiologische Klinik in Heidelberg University, the Deutsche Krebsforschungzentrum in Heidelberg, and the Forschungszentrum Rossendorf near Dresden.

From 1997 to 2008, a total of 450 patients were successfully treated at GSI, even long after my departure. I personally handed the project proposal for the HIT clinical machine in Heidelberg, the *raison d'être* for the entire endeavor, to Minister Rütgers during the inauguration of the GSI pilot project in September 1998.

The rest is history: largely built by GSI in collaboration with Siemens, HIT was approved in 2001 and has been operating successfully since 2009, with over 9000 treatments conducted to date. My colleague J. Heintze lived for three more years after the beginning of the project, allowing me to pass his test successfully. Looking back on the project, it was likely the most valuable contribution I made in my life. Beyond its scientific and technical impact, it also gave GSI a revitalized image, undoubtedly paving the way for its future.

The Journey toward the Heidelberg Ion Beam Therapy Center (HIT)

In 2001, the Science Council approved funding for the Heidelberg Ion-Beam Therapy Center (HIT), and just 3 years later, in May 2004, the foundation stone was laid. Following the completion of construction, scientific trials began in 2008. A year later, after securing the necessary medical operating license, HIT was officially inaugurated by Günther Oettinger, the Minister President of Baden-Wüttemberg. In November 2011, Minister President Winfried Kretschmann designated HIT as a "Selected Landmark in the Land of Ideas." To date, more than 9000 patients have benefited from ion-beam therapy at HIT.

Under the leadership of Hartmut Eickhoff and in collaboration with Siemens, GSI was responsible for the development and construction of HIT. This partnership led to the creation of the world's first and only heavy-ion gantry, including the fast change of different ion beams: protons (p), helium-4 (^{4}He), carbon-12 (^{12}C), and oxygen-16 (^{16}O). The Medical Director of the Radiological University Clinic Heidelberg, J. Debus, has held this role since 2003, succeeding M. Wannenmacher, while T. Haberer serves as the Technical Director (Figs. 3.85 and 3.86).

In 2015, a second clinical facility, the Marburg Ion Beam Therapy Center (MIT) at the University Hospital Marburg-Gießen, went into operation. MIT was realized by SIEMENS AG using licenses from GSI and operated under the technical and medical leadership of HIT. Approximately 2200 patients have been treated to date.

Looking Forward: The Continuing Impact of Ion-Beam Therapy

Even after stepping down as GSI's Scientific Managing Director, I remained deeply committed to advocating the expansion of ion-beam therapy. The opportunity to contribute to this field once again, 20 years later, feels like an unexpected irony of fate. Together with Ugo Amaldi, I became fully engaged in the South East European International Institute for Sustainable Technologies (SEEIIST; www.seeiist.eu). This initiative aims to extend proton and ion-beam therapy to regions where it is not yet accessible. I firmly believe that such transformative technologies should be made available to as many people as possible, regardless of geographic or economic barriers.

As Sanja Damjanovic recalls: "When I became the Minister of Science of Montenegro and initiated the development of a large-scale research infrastructure for South-East Europe, Hans J. Specht was my main supporter. In early 2017, he suggested that SEEIIST should be an accelerator-based research infrastructure for cancer therapy and biomedical research with ion beams. This choice for the project provided a significant boost and pan-European flag, garnering political support

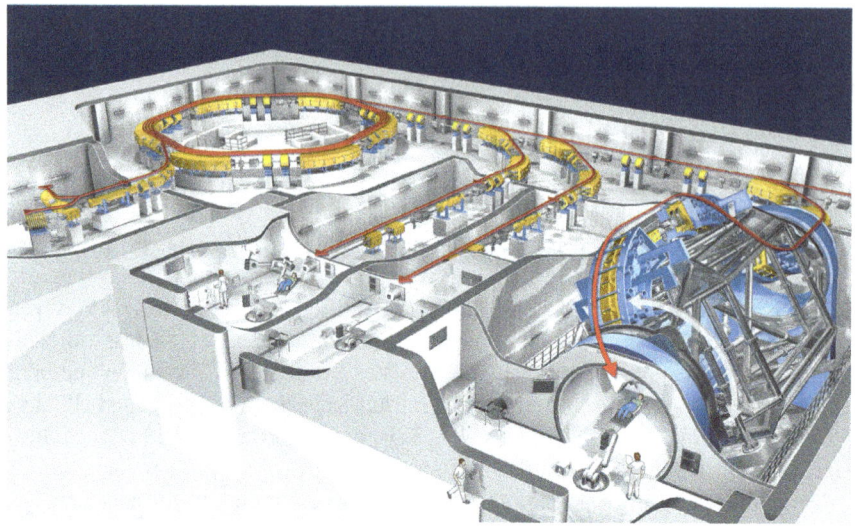

Fig. 3.85 A spin-off of GSI: the Heidelberg Ion-Beam Therapy Center (HIT). © Heidelberg Ion Beam Therapy Center (HIT). All rights reserved

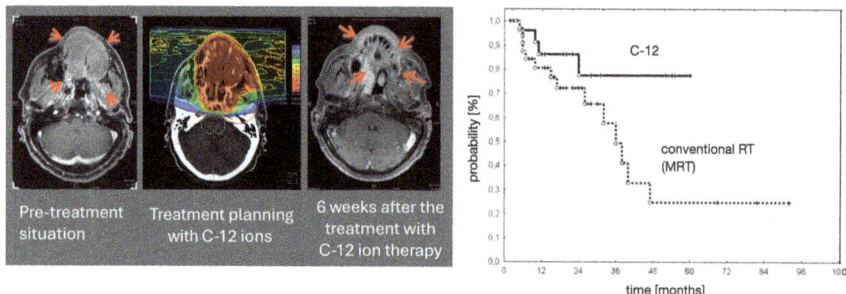

Fig. 3.86 Carbon ion therapy shows high effectiveness and survival probability. For example, in the case of salivary gland tumors, the survival probability after five years of treatment has reached 80% with carbon ions, compared to 20% with conventional radiation treatments. © Heidelberg Ion Beam Therapy Center (HIT). All rights reserved and © D. Schulz-Ertner, Cancer. 2005 Jul 15; 104(2): 338–44. All rights reserved. *Source* HIT Heidelberg

and capturing the interest of the European research community and the European Commission. Together with Ugo Amaldi, Hans J. Specht played a crucial role in developing and driving the SEEIIST project to success. He was a strong advocate for international cooperation and, for the first time, brought together four European hadron therapy cancer clinics to work together, laying the foundation for SEEIIST as the fifth center. If SEEIIST reaches full realization, it should be dedicated to Hans J. Specht and Ugo Amaldi."

Expanding Applied Research with the PHELIX Project

Another important milestone in GSI's pursuit of applied research was the development of the Petawatt-High Energy Laser for Heavy-Ion Experiments (PHELIX). This project marked a major leap forward, bridging atomic, nuclear, and plasma physics in impactful ways. PHELIX was conceived as a short-pulse, high-power laser system capable of producing laser intensities in the petawatt range, thereby enabling unique experiments harnessing the synergy obtained by combining intense laser light with heavy-ion beams. This capability created a new frontier in the exploration of extreme states of matter, such as those at very high temperatures and pressures, laying the groundwork for future applications in energy research and astrophysics.

The PHELIX project also led to my extended visits to the Lawrence Livermore National Laboratory. The collaboration enriched the PHELIX project, enabling GSI to incorporate best practices and cutting-edge technologies, further strengthening its research capabilities.

The PHELIX project proposal was completed in December 1998, and construction began in 2000. It is highly gratifying to see that this project has become a resounding success, now in great demand by a large user community, adding yet another powerful dimension to GSI's applied research portfolio alongside hadron cancer therapy (Fig. 3.87).

Fig. 3.87 The proposal of the PHELIX project was completed in December 1998; Construction officially started in 2000 with a groundbreaking ceremony attended by the Federal Minister of Education and Research, Edelgard Bulmahn. © FAIR-GSI, with a photo credit to A. Zschau. All rights reserved

Further Medium-Term Oriented Projects and Contributions to High-Energy Physics

GSI's commitment to advancing research continued with the ongoing success of the UNILAC and SIS18 facilities, which had just been completed under my predecessor P. Kienle. Beyond the study of superheavy elements, a great many significant results were achieved on both fronts. The puzzle of the infamous "GSI positrons," previously a topic of debate, was quickly resolved as an artifact, thanks to increased communication among research groups and adequate beam time allocation.

In addition to these achievements, GSI has pursued medium-term, future-oriented projects in both fundamental and applied physics. These include the SIS18 experiment HADES (High Acceptance Di-Electron Spectrometer), and GSI's participation in the upcoming ALICE experiment at the LHC.

The SIS18-based HADES experiment, approved in 1994 and completed by 1998, was optimized for electron-pair measurement, complementing GSI's existing hadron research and strengthening its contributions to high-energy physics. With the HADES project, GSI advanced its in-house experimental capabilities and strengthened collaborative ties with CERN, supporting a new research era at GSI, particularly in electron-pair spectroscopy.

Participation in ALICE at the LHC brought logistical and financial challenges which were overcome with the support of the BMFT. GSI made substantial contributions to the ALICE detectors, and its accelerator team contributed significantly to the new Pb injector at CERN using the IH accelerator structures for the linac part. This collaboration consolidated the longstanding network of relations between GSI and CERN, which continues to thrive today.

The Long-Term Future of GSI

Despite the SIS18 accelerator only becoming operational in 1990/91, I initiated discussions about GSI's long-term future, including plans for new accelerator facilities. These discussions ultimately paved the way for the development of the FAIR (Facility for Antiproton and Ion Research) project. In early 1996, we launched a structured process to outline GSI's future, forming nine internationally composed working groups with European participation. Their task was to identify promising research directions and evaluate the physics potential of possible future projects, assessing their competitiveness with regard to other plans worldwide.

These working groups focused on various fields, including deep-inelastic electron–nucleon and electron–nucleus scattering; nuclear collisions at maximum baryon density; physics with secondary beams of pions, kaons, and antiprotons; X-ray spectroscopy and radiation physics; nuclear structure physics with radioactive beams;

plasma physics with intense heavy-ion beams; conceptual design study of an electron–nucleon/nucleus collider for high luminosities; high power laser applications; and accelerator studies.

Corresponding to the different topics discussed by these working groups, two basic accelerator scenarios were developed for a possible major upgrade of the GSI facilities. The first scenario focused on a further increase in intensity, including the option of bunching for extremely high pulse currents, while the second considered an increase in proton/heavy-ion energy in combination with high-energy electron cooling and a colliding electron beam.

The aim was to arrive at a recommendation by the early summer of 1997 about which option should be pursued in order to make a serious proposal for a new project. It was clear that any such proposal should be complementary to other projects, competitive, and gain the unanimous support of the national and European science community.

Core Recommendations

The spirit of the recommendations after a 3-year discussion period, as put forth in the Status Report for GSI's Scientific Council in May 1999, can be gleaned from the following quote: "Heavy ion beams should remain the backbone of any future program, focusing on very high beam intensities rather than much higher energies; this approach aligns with the best tradition of the laboratory, ensuring uniqueness on a world-wide scale in the longer run; the present proposal so far abstains from "foreign terrain," i.e., leptonic and hadronic probes (groups 1 and 4, respectively): the first was transferred to DESY (joint workshop in Seeheim in 1997), the second (primarily antiprotons) was being discussed by a community outside GSI, mostly former LEAR users, and is the subject of their own separate document."

The final proposal recommended a two-phase project:

- **Phase 1**: A new fragment separator, an intermediate collector ring with stochastic cooling, an upgraded ESR with electron cooling, and a new electron storage ring *(filled by a new 100 MeV electron linac) for a colliding mode with fragments in the ESR*, where the new fragmentation and storage ring facility has a net luminosity gain by a factor of about 100.
- **Phase 2**: A larger 50 Tm synchrotron in combination with an 18 Tm accumulator ring, to fully leverage the infrastructure of Phase 1 with further significant luminosity gains.

By the end of 1998, the following conclusions were drawn:

(1) To forgo "foreign terrain" where corresponding capabilities at DESY and/or CERN looked as thought they would be superior in the long run.
(2) To concentrate on GSI's core areas with global uniqueness: heavy ions, a new fragment separator, improved storage rings, an electron storage ring, a synchrotron with 50 Tm or more, an accumulator ring, and achievement of the highest luminosities.

While there was an opportunity to realize at least important milestones by around 2005, in 1999, the GSI Scientific Council, in good faith, recommended deferring the project to my successor. The rest is history. As of 2017, the projection is that the first usable beam from the 100 Tm SIS100 synchrotron, now a core component of the FAIR project, is anticipated by 2025, 20 years later.

Selected Publications

- GSI Scientific Reports (1992–1999) (not accessible online)

- "Einrichtung einer experimentallen Strahlentherapie bei der Gesellschaft fuer Schwerionenforschung Darmstadt," GSI-Report 93-23, May 1993, ISSN 017145546

- GSI Scientific Report 1997, ISSN 0174-0814 (not accessible online)

  A report detailing the first patient treatments conducted as part of the heavy-ion therapy project at GSI, marking a significant milestone in the application of ion beam therapy for cancer treatment.

- GSI Scientific Report 97–05, Long Range Perspectives of GSI (not accessible online)

- Proposal for a dedicated ion-beam facility for cancer therapy (HIT proposal), 1998

- PHELIX Proposal, GSI-98–10, Report December 1998

- Status Report for the Scientific Council of GSI, May 1999 (not accessible online)

Summary of Management Milestones GSI 1992–1999 taken from Hans J. Specht's Webpage

Directorate Structure (similar to CERN, appointments also from outside)

1992 Chairman: Professor Hans J. Specht (WGF)
 Research: Professor V. Metag (Giessen),
 since 1999 Professor J. Kluge (Mainz)
 Accelerator: Dr. Norbert Angert

Technical Infrastructure: Dr. W. von Rueden (CERN)
Administration: Dr. Helmut Zeittraeger (KGF)
Scientific Secretary: Dr. K.D. Gross.

Future Internationalization of GSI

1992 Change of language in the GSI Scientific Council from German to English
Much broader level of outside advice, accompanied by a further increase in mutual lab exchanges and internationalization of the user community.

Cancer Therapy with Ion Beams

1993 Project Proposal, May 1993
"Errichtung einer experimentellen Strahlentherapie bei der GSI Darmstadt"
Radiologische Klinik der Universität Heidelberg,
GSI Darmstadt (Project lead), DKFZ Heidelberg
Successful treatment of about 450 patients from 1997–2008 in the medical cave at the GSI synchrotron SIS18. Prerequisite around 2001 for the smooth approval procedure of the clinical machine in Heidelberg.

1998 Project Proposal, September 1998
"Construction of a Clinical Therapy Facility for Cancer Treatment with Ion Beams"
Radiologische Klinik der Universität Heidelberg (Project lead),
GSI Darmstadt (accelerator), DKFZ Heidelberg (treatment planning).
*The clinical machine in Heidelberg (**HIT**) was essentially built by GSI and inaugurated in November 2009. More than 8000 patients had been successfully treated by 2024.*

ALICE Experiment at CERN

1994 Approval for participation of GSI, including funding by GSI, and funding of the German University Groups by the BMFT.

HADES Experiment at GSI

1994 Approval as the **H**igh **A**cceptance **D**i-**E**lectron **S**pectrometer
Unusually long delays in construction. Quite successful in topics of hadron production (in particular strangeness), with interesting results on di-electron emission in heavy-ion collisions emerging more recently.

Long-Term Future of GSI

1996 Establishment of nine working groups with European participation to guide future research directions and new accelerator facilities.

- Deep-inelastic electron–nucleon and electron–nucleus scattering
- X-ray spectroscopy and radiation physics
- Nuclear collisions at maximum baryon density
- Physics of secondary beams
- Nuclear structure with radioactive beams

- Plasma physics with heavy-ion beams
- Accelerator studies (electron–nucleon/nucleus collider)
- Accelerator studies (high intensity option)
- High-power lasers at GSI

1999 **Long-Term Plan for Upgrading the GSI Facilities**
(Status Report for the Scientific Council of GSI, May 1999; not public)
1999: Development of a concrete proposal in two phases: (1) new fragment separator; intermediate collector ring with stochastic cooling; much improved ESR with electron cooling; new electron storage ring (filled by a new 100 MeV electron linac) for a colliding mode with fragments in the ESR; net luminosity gain of the new fragmentation and storage ring facility by a factor of about 100. (2) new 50 Tm synchrotron in combination with a new 18 Tm accumulator ring to get the full benefits of (1) together with further significant increases in luminosity.

High-Power Laser at GSI

1998 **P**etawatt **H**igh-**E**nergy **L**aser for Heavy-**I**on E**X**periments (**PHELIX**)
Launch of the PHELIX project, a kilojoule, petawatt laser at GSI, offering stand-alone possibilities among the leading European facilities, but with the combination of high-current heavy-ion beams and intense laser beams, it provides world-wide unique synergies for basic issues in nuclear, atomic, and plasma physics, while remaining directly relevant for applied physics topics like inertial confinement fusion. Highly successful and in great demand by a large user community.

Hans J. Specht's tenure as Scientific Managing Director of GSI was marked by his bold leadership, pioneering reforms, and transformative innovations which pushed the boundaries of both fundamental and applied science. With his vision and a deep understanding of physics, he successfully navigated the challenges of managing a diverse, world-class research institution, while driving transformative initiatives such as ion-beam cancer therapy. Hans J. Specht's strategic thinking and forward-looking approach bridged the gap between cutting-edge fundamental research and high-level practical applications, elevated GSI's global standing, and laid the groundwork for the development of next-generation accelerator facilities.

As Norbert Angert recalls: "The transfer of the successful experimental therapy to industry was also decisive for a positive assessment of the proposal of the new facility FAIR (Facility for Antiproton and Ion Research). It is to Hans Specht's great credit that the foundations for this future perspective of GSI for the next generations of researchers were laid during his term of office."

Hans J. Specht's legacy extends beyond GSI, ensuring that hadron cancer therapy became a cornerstone in Europe's fight against cancer, starting with the establishment of the first European hadron cancer therapy clinic, HIT in Heidelberg—a GSI spinoff. His diverse contributions will continue to shape scientific discovery, sustainable development, and human health for generations to come.

Epilogue

For Hans J. Specht, science was a boundless source of joy and inspiration—an unending journey of discovery that shaped his diverse contributions to physics. His scientific journey was shaped by two principles he often emphasized: the profound importance of developing new methods to continually gain fresh insights in experimental science, and the critical role of teamwork at every stage of a scientific career. These principles were encapsulated in the philosophy of his mentor, H. Maier-Leibnitz, inspired by Lichtenberg's maxim: "Do something new to see something new."

Throughout his career, Hans J. Specht demonstrated an extraordinary ability to think beyond the traditional boundaries of science, transforming fundamental research into a service to society. By weaving physics into fields like music and neuroscience, he showed how science is not confined to the laboratory—it is a universal language that enriches every aspect of life. His passion for discovery—an interplay between precision, curiosity, and unpredictability—combined with his critical guidance and thought-provoking questions, enabled him to inspire countless colleagues and students, mentoring many diploma and Ph.D. candidates along the way.

Hans J. Specht's legacy is not only defined by his groundbreaking discoveries in atomic physics, nuclear fission, and high-energy nuclear physics, or by his brilliance as an experimentalist with sharp insight into cutting-edge detector technologies, but also by his outstanding character and the spirit of collaboration and openness that he brought to science. He understood that impactful science requires more than intellectual rigor—it demands humility, vision, and the wisdom to inspire teamwork and make decisions that drive progress. He valued the entire process—the meticulous experiments, the discussions, and the occasional setbacks—just as much as the breakthroughs.

As he often remarked, there was never a moment in his personal or professional life when he felt bored—a testament to his playful approach to science and his relentless drive to explore and learn.

If there is one lesson Hans J. Specht would hope to pass on to future generations of scientists, it would be this: to embrace the unknown with joy, unleashing the enduring power of curiosity and exploring the diversity of science.

Hans J. Specht's life reminds us that science is more than a pursuit—it is a joyful journey, one that enriches, connects, and transcends generations, and carries a boundless potential to change the world.

Our fascination with diversity has never faded. Reflecting on what has truly defined our lives as physicists, Hans Specht quoted J. Heintze in his emeritus lecture: "… it was indeed fun, I can assure you of that" and added: "If you asked me, I would do everything again and no differently."

Open Access This chapter is licensed under the terms of the Creative Commons Attribution 4.0 International License (http://creativecommons.org/licenses/by/4.0/), which permits use, sharing, adaptation, distribution and reproduction in any medium or format, as long as you give appropriate credit to the original author(s) and the source, provide a link to the Creative Commons license and indicate if changes were made.

The images or other third party material in this chapter are included in the chapter's Creative Commons license, unless indicated otherwise in a credit line to the material. If material is not included in the chapter's Creative Commons license and your intended use is not permitted by statutory regulation or exceeds the permitted use, you will need to obtain permission directly from the copyright holder.

Chapter 4
Reflections and Tributes

Peter Armbruster, Ewald Konecny, Volker Metag, Reinhold Schuch,
Horst Schmidt-Böcking, Helmut Satz, Jurgen Schukraft, Michael Albrow,
Axel Drees, Jochen Wambach, Wolfram Weise, Charles Gale,
Dinesh Kumar Srivastava, Gianluca Usai, Ralf Rapp, Volker Koch,
Hans Günter Dosch, Peter Schneider, Ulrich Charisius, Hartmut Eickhoff,
Thomas Kühl, Andreas Tauschwitz, Klaus Dieter Gross,
and Sanja Damjanovic

P. Armbruster (✉) · H. Eickhoff · T. Kühl · A. Tauschwitz · K. D. Gross · S. Damjanovic
GSI Helmholtz Centre for Heavy Ion Research, Darmstadt, Germany

H. Eickhoff
e-mail: H.Eickhoff@gsi.de

T. Kühl
e-mail: T.Kuehl@gsi.de

A. Tauschwitz
e-mail: A.Tauschwitz@gsi.de

K. D. Gross
e-mail: K.D.Gross@gsi.de

S. Damjanovic
e-mail: S.Damjanovic@gsi.de

E. Konecny
University of Lübeck, Lübeck, Germany

V. Metag
II. Physikalisches Institut, Justus-Liebig-Universität Giessen, Giessen, Germany
e-mail: volker.metag@exp2.physik.uni-giessen.de

R. Schuch
Physics Department, Stockholm University, AlbaNova, Stockholm, Sweden
e-mail: schuch@fysik.su.se

H. Schmidt-Böcking
Institute for Nuclear Physics, University Frankfurt, Frankfurt, Germany
e-mail: schmidtb@atom.uni-frankfurt.de

© The Author(s) 2025
H. J. Specht et al. (eds.), *Hans Joachim Specht*, Springer Biographies,
https://doi.org/10.1007/978-3-031-92353-1_4

H. Satz
Faculty for Physics, University of Bielefeld, Bielefeld, Germany
e-mail: satz@physik.uni-bielefeld.de

J. Schukraft
CERN, Geneva, Switzerland
e-mail: Jurgen.Schukraft@cern.ch

M. Albrow
Fermi National Accelerator Laboratory, Batavia, IL, USA
e-mail: albrow@fnal.gov

A. Drees
Department of Physics and Astronomy, Stony Brook University, Stony Brook, NY, USA
e-mail: axel.drees@stonybrook.edu

J. Wambach
Institute für Kernphysik, Technische Universität Darmstadt, Darmstadt, Germany

Department of Physics, University of Illinois Urbana-Champaign, Urbana, IL, USA
e-mail: jochen.wambach@tu-darmstadt.de

W. Weise
Physics Department, Technical University of Munich, Garching, Germany
e-mail: weise@tum.de

C. Gale
McGill University, Montreal, QC, Canada
e-mail: charles.gale@mcgill.ca

D. K. Srivastava
National Institute of Advanced Studies, Bengaluru, India
e-mail: dinesh.srivastava@nias.res.in

G. Usai
University of Cagliari, Cittadella Universitaria, Monserrato, Cagliari, Italy
e-mail: gianluca.usai@ca.infn.it

R. Rapp
Department of Physics and Astronomy and Cyclotron Institute, Texas A&M University, College Station, TX, USA
e-mail: rapp@comp.tamu.edu

V. Koch
Nuclear Science Division, Lawrence Berkeley National Laboratory, Berkeley, CA, USA
e-mail: vkoch@lbl.gov

H. G. Dosch
Institut für Theoretische Physik, Universität Heidelberg, Heidelberg, Germany
e-mail: h.g.dosch@gmail.com

Early Research Years and Our Collaboration at the FRM Reactor in Munich

Peter Armbruster

A Career Starts with a Thin Window Foil

It was in the year 1960 that our paths first crossed. Hans was looking for a diploma thesis with Professor Maier-Leibnitz, and I was fortunate to have him assigned to me, an assistant at the Institute of Technical Physics at the Technical University of Munich (TUM). I had planned to conduct experiments on the beta decay of radioactive isotopes produced by fission with thermal neutrons, using the gas-filled separator at the Garching reactor, one of the first recoil mass spectrometers designed to separate fission products by mass.

Our main challenge was distinguishing beta-delayed gamma radiation in the beta decay chains from the beta particles we wanted to detect. This was achieved by setting up an anti-coincidence system between signals from a thin CH_4-filled counting tube and a thick scintillator, which measured the residual energy of the beta particles. Hans's task for his diploma thesis was to construct this counting tube.

Hans completed the task in an incredibly short time, to everyone's satisfaction. The issue of the charge polarization of the primary fission fragments—which we had aimed to solve—was particularly challenging, and it was a problem we revisited at the FRS. The counter Hans built played a crucial role in helping our colleague Herbert Meister to complete his doctoral thesis, and it also led to Hans's first scientific publication as a diploma student.

Hans's doctoral thesis continued my own work: an experiment focused on the excitation of inner electron shells in heavy-ion collisions, specifically within the mass range of fission products at energies around 1 MeV/u. It was, as we would call it today, a secondary beam experiment in atomic physics. At that time, thin self-supporting foils were scarce, so I challenged Hans to find a solution.

Although thick foils were readily available on the market, they posed a problem, as fission products could get stuck in them, preventing the generation of coincidence signals between the ions and X-ray photons induced by collisions. Once again, the solution was a transmission counter tube, though this time not for beta particles, but for heavy ions.

This counter chamber, with thin windows and filled with a few Torr of CH_4, consisted of a three-chamber counter tube with partition walls. It was capable of

---

P. Schneider
Institute of Psychology, "Music Psychology Und Brain Research" Section, University Graz, Graz, Austria
e-mail: p.schneider.hd@web.de

U. Charisius
Heidelberg, Germany
e-mail: ulrich@charisius.de

operating with or without gas amplification. Hans was deeply engaged in studying the function of this counter tube. He discovered that, by adjusting the counter tube voltage and gas pressure, he could cover the entire range from the onset of detection to high gas amplification. This achievement, although not fully recognized at the time, was a pioneering effort. *Had Hans removed the walls, it would have become the first multi-wire chamber for heavy ions.*

The study of inner-shell excitation proved to be a lasting success. Resonances in the dependence of the cross section on atomic number were explained using two-center models of the atomic levels of the collision system. When the nuclear charges of the collision partners are in certain ratios, electrons can be promoted to higher shells of the quasi-molecule, resulting in resonances in the cross section. This mechanism, in which electrons transition down to the 1s states of the united system, leads to positron production—a phenomenon later discovered in heavy-ion collisions at GSI.

In 1964, the year of my habilitation and Hans's Ph.D., I moved to KFA Jülich, and shortly thereafter, Hans went to Chalk River. Hans's exceptional skill in handling thin window foils impressed us all. As a result, Hans was tasked with building all the detectors and chambers—no one could match his precision.

I also remember that Hans slept very little. The reactor ran day and night, and Hans was always there at the experiment. It wasn't public transport to Garching that made this possible, but rather a little green car we all admired. My son, now almost 40 years old, still talks fondly about it. Those were good times for us, even as the world teetered on the brink during the Cuban Missile Crisis (Fig. 4.1).

Fig. 4.1 From right to left: Peter Armbruster and Hans J. Specht, GSI, April 2021. The association between Peter Armbruster and Hans J. Specht was characterized by deep scientific respect and genuine personal warmth, and endured to the very end. © FAIR-GSI, with a photo credit to G. Otto. All rights reserved

Neutron Evaporation and Other (Mis)Deeds of the Young Specht from the Munich Years

Ewald Konecny

Life was busy, beautiful, and cheerful during those early days of building the Munich Tandem Accelerator in the early 1970s. We worked hard but celebrated our successes with equal fervor.

While searching for conversion electrons in the "second minimum," we endured intense beam time periods which often stretched continuously over 3 weeks, with only four of us managing the experiments. One memorable incident involved assigning a team member to monitor neutron evaporation events in the dimly lit accelerator hall. Hans was tasked with reporting these observations via walkie-talkie. After several minutes without a response, Hans went to check and found the observer peacefully asleep on the hall floor.

On the other hand, each multi-coincidence event, which occurred roughly every 15 min, was met with an acoustic signal, joyfully acknowledged by the team. Back then, our celebratory signals were modest piezoelectric beeps, but today they would likely be accompanied by a grand multimedia fanfare worthy of Wagner.

Even our smaller successes garnered enthusiasm as we celebrated along the Garching Isar meadows with bonfires. Triumphs were marked not just by the wooden cores of cable spools, but also by the construction workers' latrines, which were inadvertently transformed into glowing gas. This, of course, led to warning letters from the university administration, prompting the installation of metal latrines.

Larger experiments naturally demanded careful logistical coordination across various institutes. The young Hans quickly recognized the significant influence of secretaries in managing complex projects. He adeptly charmed them, especially those who were strikingly feminine and graceful, receiving eager cooperation in return.

When it came to handling radioactive targets, Hans preferred to rely on his own touch. Even the thinnest surgical gloves felt inadequate, leading him to manipulate targets with his bare hands—resulting in what we humorously referred to as his "radiant" fingers. One incident during the preparation of an actinium target involved a radium ampoule breaking and spilling its contents across the laboratory floor. While his co-experimenter entertained the arriving firefighters with beer and pretzels, Hans remained focused on finishing the target preparation, ensuring the experiment commenced with only a slight delay.

This setting became pivotal for our research on the "second minimum." The accelerator operated at full capacity, utilizing nearly every available transuranium element as targets. Thanks to an excellent target laboratory and strong connections to Los Alamos for raw material procurement, we had ample resources to work with. However, bureaucratic processes struggled to keep pace with Hans's relentless research enthusiasm.

For instance, material had to be obtained from another ministry, which classified our needs as "only" radioactive materials. While we had the necessary permission to work with ^{235}U and ^{233}U, the approval for ^{234}U had yet to arrive. Despite their identical chemistry and closely related radioactivity, bureaucratic rules mandated strict adherence, making it challenging to navigate our requirements. To ensure the integrity of our meticulously planned experiment, we resorted to relabeling the ^{234}U we intended to process as ^{235}U. In a sense, the extra neutron "evaporated," allowing us to proceed.

Hans J. Specht's contributions to nuclear fission and his innovative spirit significantly shaped our understanding of nuclear processes. His legacy, woven through laughter and dedication, remains a testament to the collaborative spirit which defined our research community.

Impressions from the Heidelberg Tandem Accelerator and Later Encounters with Hans Specht

Volker Metag

When we were overtired after the night shift, relieved that the first data had been written to magnetic tape and that the online analysis provided reasonable-looking spectra, Hans would arrive to check the electronics settings. "Who adjusted this?" he would exclaim, as the little yellow screwdriver was swiftly drawn out, his glasses removed so he could observe even minute signal shifts on the oscilloscope. Trigger thresholds and signal delays would be tweaked, and the supposedly fine-tuned fast/slow coincidence branches optimized once again. This ritual was a hallmark of every beam time at the Heidelberg Tandem Accelerator.

These experiments yielded new insights into the spectroscopy of vibrational states in the second minimum of the nuclear potential and their coupling to the fission degree of freedom, as well as to states in the first minimum. This topic had been first explored by Hans during his time at Chalk River, in collaboration with J.C.D. Milton and J.S. Fraser, and was continued at the Munich Q3D spectrograph after his return.

But there were also experiments that didn't involve much electronics. A series of experiments was based on the "charge-plunger technique." This method combined Dieter Habs' ingenious idea of resetting the high atomic charge states of fission isomeric recoils—caused by conversion transitions in the second minimum—when crossing thin carbon films with Volker's extensive experience in fission isomer lifetime measurements using recoil-distance and projection methods, the production of planar actinide targets, and fission fragment detection using track detectors. These experiments were not without incident. On one occasion, when we vented the scattering chamber after three days of beam time, we discovered that the actinide target had burst. The timing of the burst could not be determined, rendering the measurement unusable. Worse still, the scattering chamber was contaminated. Under the strict supervision of Mr. Festag, the radiation protection officer, Dieter, myself, Günther

Ulfert, and Helga Krieger worked diligently, cleaning the chamber until it was spotless and free of detectable radioactivity. Hans stopped by, watching the masked team at work, and offered the comforting words, "Rome wasn't built in a day either." Despite the setbacks, we persevered and ultimately succeeded. In 1978, Dieter and Volker were awarded the Physics Prize by the German Physical Society for their work on determining the quadrupole moments of fission isomers using the "charge-plunger" technique, a significant milestone that paved the way for our later academic careers.

Discussions of our results were always lively and focused. Only a few things could distract Hans—dark-eyed female doctoral students or heavy downpours. Once, during a rainstorm, he darted down the stairs in the Bothe lab as though he had been bitten by a tarantula. The Lotus parked in the no-parking zone (reserved for MPI directors) had flooded, and Hans immediately began scooping water out to prevent further damage.

The culmination of our seven-year collaboration was the Physics Reports article (C65, 1980), which summarized all the results from this highly productive and intellectually stimulating period.

Hans and I didn't just work together at the Heidelberg Tandem but also later for a year at CERN, combining the CERES and TAPS detector systems to study Dalitz and dilepton decays of neutral mesons in ultra-relativistic, proton-induced reactions. Later, from 1993 to 1998, we worked together in the directorate at GSI, where Hans as Director General appointed me as Research Director. After the start of the UNILAC program in the 1970s and the extension with the SIS18 synchrotron and the experimental storage ring (ESR) in the late 1980s, GSI had evolved into a leading international research center for heavy-ion beams. To broaden the scientific scope of the facility, we began extensive discussions on future research opportunities, aiming to offer unique capabilities that would complement other facilities. Eight international expert working groups were established to explore potential research areas, such as an electron–ion collider, physics with intense secondary hadron beams (including antiprotons) and beams of radioactive exotic nuclei, atomic physics in extreme fields, plasma physics with intense heavy-ion beams, and accelerator studies. My role was to oversee these discussions and summarize the recommendations, which were then debated in a larger workshop involving all the groups. The proposal for an electron–ion collider was particularly well received. However, when Björn Wiek, director of DESY, remarked, "If it's that interesting, we'll do it at DESY," Hans accepted the statement without hesitation, much to my disappointment as I had been one of the conveners of the working group together with Dietrich von Harrach and Andreas Schäfer, advocating for this project. This concept is now being pursued at BNL, at much higher and more suitable energies for studying the internal structure of hadrons. Nevertheless, the working groups' recommendations laid the foundation for the decisions that eventually led to the development of the FAIR project, which is currently under construction.

In 2006, the University of Giessen celebrated its 400th anniversary with a series of special events. Normally, the physics colloquium is attended by about 100 people, but on 7 May, we had to move the event to a much larger lecture hall, seating

450. That day, I saw many unfamiliar faces in the audience. A piano and several detector systems had been set up on stage, and the colloquium began with a string quartet from the university orchestra. Then Hans and Hans Günther Dosch arrived. I had invited them to deliver their famous lecture *Musical Perception—Physics, Neurophysiology, Psychology* with experiments and musical demonstrations. This lecture, already presented at numerous prestigious institutions, captivated the audience, sparking many questions. It turned into the longest physics colloquium I had ever attended.

In 2007, Hans Specht and I attended the International Nuclear Physics Conference in Tokyo, organized by Shoji Nagamiya. A special highlight of the conference was a reception with the Japanese Emperor. A selected number of participants, including myself, were allowed to speak to the Emperor, each for about 20 s. We stood in a line and waited for our turn. Hans was two people in front of me. As soon as Hans had been allowed to speak to the Emperor, the reception ended due to time constraints. My conclusion was: "Hans had the last word—as always."

In 2010, at the Chiral10 conference in Valencia, my colleague Mariana Nanova and I were heading to the hotel after a full day of talks. As we discussed where to have dinner, suddenly someone approached us with papers full of notes—Hans! He had attended both of our presentations and had a list of follow-up questions that had been cut short by the strict session chair. We continued our discussion on resonance widths on the street, and dinner was delayed by 2 h. This was typical of Hans: whenever he spotted a weak point in the argument of a presentation, he immediately sought to clarify it.

Although working with Hans was sometimes challenging, I was always deeply impressed by his unwavering commitment to uncovering scientific truth, his intellectual rigor, and his remarkable rhetorical skills. His breadth of knowledge and wide-ranging interests left a lasting impact on everyone he worked with (Fig. 4.2).

Fig. 4.2 Volker Metag (GSI research director) at the presentation of a book with joint publications and contributions by collaborators on the occasion of Hans Specht's 60th birthday. © GSI/FAIR, with a photo credit to A. Zschau. All rights reserved

Hans J. Specht's Ph.D. Thesis: Opening a New Field of Quasi-atoms

Reinhold Schuch

It was in the late spring of 1974 when I first came into contact with Hans Joachim Specht at the MPI for Nuclear Physics in Heidelberg. He had just moved from Munich to take up his chair position at the University of Heidelberg, whereas I had just completed my Ph.D. exam at the University of Göttingen. I was looking for new opportunities and a change in my scientific direction, and Specht was still curious to find out more about the discoveries he had made during his Ph.D. studies 10 years before. Specht was still very enthusiastic about his findings of huge peaks in the characteristic X-ray production cross section by fission products at the Munich reactor. He had realized that he had opened a new field for heavy-ion atomic physics, one that was worthwhile to continue. He himself had moved into nuclear physics and had already been captured by the second minimum of fission. Specht saw this as an opportunity for me to start a career in physics by delving deeper into this new branch of atomic physics. In particular, the presence of the tandem accelerators at the MPI for Nuclear Physics in Heidelberg, and later at the newly founded GSI in Darmstadt, made this opportunity even more attractive.

Hans Joachim Specht began his Ph.D. thesis in 1962 in the legendary department of Heinz Maier-Leibnitz at the Technical University Munich, using the FRM reactor. At the same time, Peter Armbruster, who had just finished his Ph.D. with a thesis titled "*Bau eines Massenseparators für Spaltprodukte und Nachweis einer Anregung innerer Elektronenschalen bei der Abbremsung von Spaltprodukten*" (Construction of a mass separator for fission products and detection of excitation of inner-shell electrons during the stopping of fission products), was also involved in the research. Armbruster's thesis was published with the title "Ionisierung innerer Elektronenschalen bei der Abbremsung von Spaltprodukten" (Ionization of inner electron shells during the stopping of fission products) in *Zeitschrift fur Physik 166, 341 (1962)*. He found surprisingly strong characteristic X-rays connected with fission induced by the neutrons from the FRM reactor in a ^{235}U target. Armbruster speculated about the origin of the L and M X-rays that he detected in slow coincidence with the ions from fission penetrating different target materials, and finally concluded that they must originate from inner-shell ionization in collisions during the slowing down of the fission products in the targets. However, at that point, he had not yet conducted a systematic study of the Z-dependence of this ionization cross section. Hans Specht then took over this project from Peter Armbruster. They improved the mass analyzer and the photon detector, and published a brief article together: P. Armbruster, E. Roeckl, H.J. Specht, and A. Vollmer, *Z. Naturforsch. 19a, 1301 (1964)* "*Die Untersuchung fast-adiabatischer Stöße mit Spaltprodukten*" (The investigation of near-adiabatic collisions with fission products). In this paper, they presented the peaks in the Z-dependent normalized L- X-ray intensity. They investigated various possible reasons for these peaks and concluded: "*Die Lage der Maxima deutet auf Elektronenaustausch-Phänomene innerhalb des quasi-molekularen Zwischenzustandes während des Stoßprozesses hin*" (The position of the maxima indicates electron exchange phenomena within the quasi-molecular intermediate state during the collision process).

By measuring the ion energy dependence of the cross section, Specht concluded that the peaks got more distinct as the ion energy decreased, and that inner-shell vacancies were effectively produced even for collision velocities much smaller than the orbital velocities of the electrons concerned. The mechanism causing the high cross sections in this velocity regime is today known to be mainly electron promotion within the quasi-molecule formed during the collision. This represents the main conclusion of Specht's Ph.D. thesis, published in *Zeitschrift für Physik 185, 301 (1965)*, titled "*Ionisation innerer Elektronenschalen bei fast-adiabatischen Stössen schwerer Ionen*" (Ionization of inner electron shells in near-adiabatic collisions of heavy ions). This paper provides a detailed discussion of the mechanism that could result in increased inner-shell ionization through quasi-molecular states transiently formed during the collision process. The arguments remain mostly qualitative, featuring molecular potential energy curves for different Z-asymmetries and a correlation diagram as a function of the internuclear distance, showing possible promotions from the K shell to the L shell, as well as from the L shell to the M shell, and so on. At this stage, without detailed theoretical models, specific details on how the excitation mechanism works could not be given. Specht was the first to present explanations for the resonance-like behavior of the cross section: one explanation involves the increased ionization probability due to a reduced binding energy in the quasi-molecule, and the other is related to curve crossings within the quasi-molecule.

In the discussion part of his 1965 paper, Specht writes: "The most important result of the present experiment is the observation of the maxima in the effective cross-section for the ionization of inner electron shells in collision partner combinations in which the binding energy of the electrons in the ionized shell of the target atom approximately matches that of the electrons in any shell of the projectile. The maxima are more pronounced the more "adiabatic" the collision is. The form of the Born approximation used here neglects the shell structure of the projectile and is therefore unusable for the collision problem at hand. The fluctuations in the ionization cross sections, superimposed on the monotonic dependencies, can only be explained by the interaction of the two electron shells." The measured properties of the maxima, such as position and width, and their behavior with decreasing relative speed of the collision partners, enable at least a qualitative interpretation of the phenomenon in the picture of quasi-molecules.

This pioneering discovery, however, was not immediately recognized internationally. A likely reason for this was that Specht's papers were written in German. Then, as is often the case, around the same time, a paper was published by the St. Petersburg group led by V.V. Afrosimov et al. (*Zh. Tekhn. Fiz. 34, 1613, 1624, 1964*). In contrast to Specht's article, this was quickly translated into English and triggered U. Fano and W. Lichten's famous 1965 paper (*Phys. Rev. Letters 14, 627, 1965*), which explained the effect of the peaks in the collision inelasticity observed by Afrosimov et al. as electron promotion to the continuum in the transiently formed quasi-molecules. In fact, there was also an explanation by Afrosimov's colleague M.Ya. Amusia (*Phys. Letters 14, 36, 1965*), who proposed that collective oscillations of the electronic shells of the colliding atoms were responsible for these inelasticity peaks. These publications by Afrosimov et al., and the differing theoretical interpretations, triggered feverish research activity.

W. Brandt referred to the mechanism behind these maxima as Pauli excitation, which occurs in low-velocity quasi-adiabatic atomic collisions (*Phys Rev Lett. 24, 1037, 1970*). This paper contains the first citation to Specht's work in a side note. Specht's paper soon gained recognition, and shortly after T.M. Kavanagh et al. from the Livermore group wrote in their article (*Phys. Rev. Letters 21, 1473, 1970*): "Cross-sections have been measured for L X-ray production for a wide range of collision partners (heavy ions and atoms) in the energy range 40 keV–1.1 MeV. We have observed a strong cyclic dependence of the cross-section on the atomic number [...] a Z dependence for X-ray production cross sections has previously been reported by Specht for the case of "light" and "heavy" fission fragments incident on a wide range of target materials. Our data will be discussed in terms of a model presented by Specht, and Fano and Lichten, in which inner-shell excitations occur through level crossings in the quasi-molecule formed during the collision."

Indeed, Specht's original explanation for this effect, as seen in his Ph.D. thesis, pointed in the right direction. It took around 20 years of further studies, including more detailed experiments and theoretical models, before the picture of collisional quasi-molecules was firmly established.

With the advent of heavy-ion accelerators that were installed in many laboratories, atomic collisions became a very lively research area, particularly in the United States, but also in Europe, with GSI at the forefront. A driving goal was to create superheavy quasi-molecules, aiming to reach a united-atom limit of Z larger than 170 in collisions of very heavy ions and atoms. A breakdown of the existence of atomic bound states for united atoms with Z > 170 was predicted, and also the spontaneous decay of the vacuum into electron–positron pairs in the extremely strong field.

In 1975, when my research activity in this field started, an avalanche of papers appeared, culminating in the 1980s. Specht had foreseen this development, although he left the field of atomic collisions to focus on nuclear shape isomers and the concept of a "double-humped" fission barrier. We were thus given complete freedom to select our research topics. Luckily, I entered into a close and very fruitful collaboration with Horst Schmidt-Böcking's group at Frankfurt University. In 1975, we started experimenting with coincidences of X-rays with the scattered ions to fix the collision trajectory, which allowed us to vary the distance of closest approach. Soon, however, we ran into the problem that surface barrier ion detectors were quickly dying in the experiments by irradiation damage. Here Specht gave us some extremely valuable advice: to develop parallel-plate avalanche counters for detecting the scattered ions. This solution proved to be the key to measuring the 10^{-5} emission probabilities of quasi-molecular radiation per collision.

After a long research journey, which provided many insights into the X-ray production mechanisms of heavy ions, further validating Specht's original idea, we finally succeeded in detecting the phase oscillations. From this, we were able to

Fig. 4.3 Discussion of our data at MPIK, Heidelberg, 1976. © R. Schuch. All rights reserved

determine the energy values of the quasi-molecules, resolved in collision times in the order of 10^{-20} s. This opened the way for plans to make spectroscopic measurements of superheavy quasi-molecules, aiming to reach a united-atom limit of more than 170 in collisions of very heavy ions and atoms. However, the limited experimental and accelerator technologies available at the end of the 1980s forced us to stop in the regime of medium-heavy ions and to change our research field. Now, with the new heavy-ion machines at FAIR in Darmstadt and HIAF in China, the study of superheavy quasi-molecules is experiencing a revival, taking Specht's idea into new territory (Fig. 4.3).

A Scientist Who Knew How to Inspire

Horst Schmidt-Böcking

I met Hans Joachim Specht for the first time in 1973, after he had taken over the directorship at the Physics Institute of the University of Heidelberg from my doctoral supervisor, Otto Haxel. As I was a scientific assistant at his institute in 1973, he became my "boss." At that time, we (Itzhak Tserruya, and from 1974 also Reinhold Schuch, as well as diploma and doctoral students) were investigating electronic

4 Reflections and Tributes

Fig. 4.4 Discussion table at the MPI für Kernphysik in 1976. From left to right: Wolfgang Lichtenberg, Horst Schmidt-Böcking, Hans Joachim Specht, and NN. © Horst Schmidt-Boecking. All rights reserved

processes in fast ion–atom collisions. Since Hans J. Specht had also dealt with the excitation of electron shells in symmetric ion–atom collisions in his doctoral thesis, he showed particular interest in our work. Although I moved to the University of Frankfurt in 1974, I continued to work with him well into 1980, which is reflected in seven joint publications (Fig. 4.4).

Hans J. Specht had a decisive lifelong influence on my subsequent research work in the field of atomic and molecular physics. In 1973, he encouraged me to build a spatially resolving detector for the detection of fast ions, with which the location and time of incidence could be determined with high resolution. This was the beginning and foundation that later allowed us to measure dynamic processes in atomic and molecular physics which were thought to be impossible to visualize. Unrecognized at the time, it was more or less the beginning of the development of the COLTRIMS reaction microscope. Why?

In 1973, at the request of Johannes Hansteen from the Niels Bohr Institute in Copenhagen, we began to investigate the impact parameter-dependent inner-shell ionization and radiation emission processes in lithium-ion–atom collisions. As was common practice in our group at the time, we used so-called "Si surface barrier detectors" for ion detection and so-called "lithium drifted Si junction detectors" for photon detection. Both detectors had very small solid angles and could not process high counting rates, so our measurements were extremely time-consuming.

Since Hans J. Specht was an expert in the field of self-made gas detectors, he suggested that we should build a parallel-plate avalanche counter filled with gas for ion detection, because such a detector could process 10,000 times higher counting rates and it could also be manufactured with spatial resolution. This meant that this detector had a 4π solid angle and all relevant collision parameters could be measured simultaneously. Gebhard Gaukler, then a graduate student in my working group, was to build this detector.

The detector principle is shown in Fig. 4.5. The blue area indicates the gas filling (operation at about 10 millibar gas pressure). The yellow areas show the spatially resolved anodes (various impact parameters), each of which was read out using separate, very fast electronics. The prototype built by Gebhard Gaukler had only three 2π ring anodes (three impact parameter ranges). It worked right away and was able to register a particle rate of approximately 1 gigahertz (GHz) without any problems. This was a breakthrough in our coincident detection technology. From now on, we could measure processes where the reaction probability was even below 10^{-9} (Fig. 4.6).

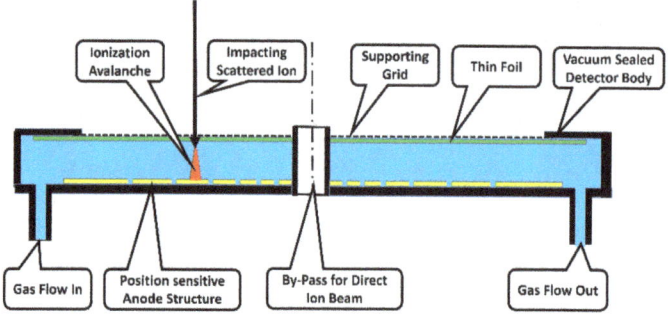

Fig. 4.5 Detector scheme (see text). © Horst Schmidt-Boecking. All rights reserved

Fig. 4.6 First prototype parallel plate avalanche counter 1973–74. © Horst Schmidt-Boecking. All rights reserved

> **Impact-Parameter Dependence of Noncharacteristic Radiation Emitted in 35-MeV Cl-Cl Collisions**
>
> I. Tserruya*
> *Max-Planck Institut für Kernphysik, Heidelberg, West Germany*
>
> and
>
> H. Schmidt-Böcking, R. Schulé, and K. Bethge
> *Institut für Kernphysik, Universität Frankfurt, Frankfurt, West Germany*
>
> and
>
> R. Schuch and H. J. Specht
> *Physikalisches Institut, Universität Heidelberg, Heidelberg, West Germany*
> (Received 19 January 1976)
>
> The impact-parameter dependence of the noncharacteristic radiation emitted in 35-MeV Cl-Cl collisions has been measured. The coincidence spectra are very similar in shape to the "singles" x-ray spectrum, and show almost no dependence on the impact parameter. In the quasimolecular-radiation picture, the quasistatic approximation is in clear disagreement with the coincidence results.

Fig. 4.7 First publication with Hans Joachim Specht. © 1976 American Physical Society. All rights reserved

The first joint publication was published in Physical Review Letters in 1976. It was the first collision parameter-dependent measurement of the so-called "quasimolecular" X-rays in ion–atom collisions (Fig. 4.7).

Following the success of this first self-built spatially resolving detector, we manufactured many different detector structures over the next three decades: gas-filled spatially and temporally resolving and energy-resolving X-ray detectors, spatially and temporally resolving neutron detectors, Xe-filled high-pressure gas scintillation X-ray detectors with fast timing properties and high Compton suppression capability, and spatially and temporally resolving detectors for very low-energy ions and electrons (milli eV). The experience gained in this research led to the founding of the company RoentDek in 1990, which today supplies detectors (with electronics) and COLTRIMS reaction microscopes to research institutes worldwide. The latter microscopes now make it possible to observe the electronic dynamics of electrons in atoms and molecules on the zeptosecond scale.

Without the stimulation initiated by Hans J. Specht in 1973 to build such special detectors, it is likely that none of our subsequent research work would have been achieved. In the following three decades, I met Hans J. Specht several times at GSI, where I was able to speak to him. I particularly remember meeting him in 2011, when he gave a lecture on music and physics in Frankfurt. On this occasion, he also visited the laboratory (Fig. 4.8) where Gerlach successfully carried out the so-called Stern–Gerlach experiment in 1922.

Fig. 4.8 Hans Joachim Specht visiting Frankfurt. This is the room where the Stern–Gerlach experiment was carried out in 1922. © Horst Schmidt-Boecking. All rights reserved

Hans J. Specht (1936–2024)

Helmut Satz

Since antiquity, an essential question for understanding nature was the composition of matter: what were its ultimate constituents? Atoms, as building blocks, were found to consist of nucleons; but in the latter half of the last century, it was shown that, according to quantum chromodynamics (QCD), these were in turn bound states of quarks. It made quark matter, rather than nuclear matter, the most fundamental form of strongly interacting matter. This conceptual transition determined physics around 1980, theoretically as well as experimentally, and Hans J. Specht's work was crucial in the evolution of this new field of physics.

On the theoretical side, the lattice simulation of QCD established the new state of matter and the quark–hadron transition. Experimentally, the idea was to study the phenomenon through high-energy nuclear collisions, and from the outset, Hans J. Specht played a leading role in the development of quark–gluon physics. It required someone who had experience in both theory and experiment. The aim was more than just extending nuclear collision experiments to higher energies: it was determining the two-state nature of strongly interacting matter, with the quark–gluon plasma as a new state. Hans came from atomic and nuclear physics, where he had already had a distinguished career, but now he entered with full force into the new field of physics, which he profoundly influenced and shaped.

Energy density estimates showed that the SPS at CERN and the AGS at Brookhaven could provide values in the region expected to lead to the transition, and so in the early 1980s, such experiments were put on the schedule, with Hans among their main proponents. Besides his general support for the research, he became the spokesperson for the experiment NA34 (HELIOS) at CERN, and subsequently followed this up there with NA45 (CERES). The aim of these experiments was to detect thermal dileptons, one of the most prominent signals of quark–gluon plasma formation. The success of the search provided one of the basic arguments in favor of CERN's subsequent claim to have produced quark–gluon plasma in the laboratory. The follow-up studies at RHIC and the LHC benefitted greatly from Hans' continued push and support.

Hans' interests had always been quite multi-faceted, but during his years as GSI director, they continued to spread. Under his guidance, the GSI became one of the pioneering research institutes introducing heavy-ion radiation for cancer therapy—one of the main areas of work there today and of great benefit to humankind in general, with several hundred successfully treated patients. Hans always emphasized that this was perhaps the work he was most proud of.

Hans had always been very much interested in music, being a gifted piano player. In his later years, together with Gunther Dosch, he investigated the neurological basis for music perception in humans, leading to several well-recognized studies.

I first encountered Hans in the early 1980s. We shared the interest in quark–gluon plasma formation and continued to remain in close contact ever after. Together with other colleagues, we launched a conference series called "Quark Matter," on deconfinement in high-energy heavy-ion collisions, and Hans already played a decisive role at the first of these meetings, in Bielefeld in 1982. In 1986, the Quark Matter Conference was held in Nordkirchen, Germany, organized by Hans, Reinhard Stock, and myself. It was the first international conference at which the new high-energy heavy-ion results were presented. Subsequently, we came together at numerous workshops, and as GSI director, Hans also did much to support theoretical progress in the field. In particular, 10 years later, in 1996, the Quark Matter meeting came back to Germany for the twelfth in the series, once again with Hans as director (Fig. 4.9).

Fig. 4.9 A meeting of experimental and theoretical physicists in ultra-relativistic heavy-ion research took place at the Quark Matter 1996 conference in Heidelberg, attended by 550 participants. From left to right: Hans J. Specht, Peter Braun-Munzinger, Helmut Satz, Johanna Stachel, and Reinhard Stock, during a press conference. © Hans J. Specht. All rights reserved

As for the Future, Your Task Is Not to Foresee, but to Enable It: The Inception of the CERN Heavy-Ion Program

Jurgen Schukraft

Antoine de Saint-Exupéry
The Wisdom of the Sands
CERN, Geneva, Spring 1983

These were busy times at CERN, the European high-energy physics Laboratory. The Intersection Storage Ring (ISR), the first large hadron collider ever to be built, and since 1971, the frontier of high-energy physics research, was scheduled for closure by the end of the year to free up resources for CERN's LEP project. The members of the Axial Field Spectrometer (AFS) Collaboration, an experiment at the ISR looking for direct photons and jets, were crammed into their small meeting room to listen to a seminar by Bob Palmer. He reported on the production of "soft," i.e., low p_T, electrons in hadron collisions, where an anomaly had been reported in a number of recent experiments, an excess apparently not related to any known source or process. Direct leptons at high p_T had been studied in hadron reactions for over 10 years and had been found to be connected to "new physics" at the time (e.g., heavy quark production, the Drell–Yan process), so any anomaly had to be taken seriously, maybe indicating "new physics" once again.

The following days were filled with heated discussions and wild speculations. Was this signal real? What could be its origin? Could there be a connection with a similar ill-understood excess observed in soft photon production? Could the AFS experiment be used to measure these electrons? In the few remaining months of operation? Maybe! Action!

4 Reflections and Tributes

Within a few weeks of unprecedented activity, the AFS was reconfigured and upgraded with equipment borrowed from other experiments or salvaged from the stocks to add electron identification. A small team assembled the new hardware, set up a dedicated electron trigger, and managed, just in the nick of time, to take data before the last colliding ISR beams were finally dumped in December of 1983.

And in the midst of this team of high-energy physicists, a newcomer to CERN, Hans Joachim Specht, a "Herr Professor" from nuclear physics, was busily estimating rates and acceptances, pulling cables, adjusting electronics, and running data-taking shifts.

Hans had arrived at CERN for a sabbatical and taken up this new venture of soft electrons (and later, photons) as *his* subject. This was a theme he would follow up, promote, and later guide over the coming years, from the first shot at the ISR to the CERN fixed target experiments HELIOS/1, HELIOS/2, and CERES (the ChErenkov Ring Electron Spectrometer), to HADES at GSI, and far into the future, ALICE at the LHC. Not forgetting the soft photon experiments SOPHY and BACY.

How come, this nuclear physicist, an expert on low-energy nuclear reactions, fission barriers, and the like, turned up at CERN in the closing days of a high-energy particle physics machine? Everything must have started around 3 years earlier, when a workshop was held at GSI in October 1980 on "Future Relativistic Ion Experiments." BEVELAC physics was then in full swing, with new large acceptance detectors in operation or under construction (streamer chamber, plastic ball), when physicists involved in this still young research field were already contemplating ways of doing "really" high-energy nuclear collisions.

A mixed group from both nuclear and high-energy physics had gathered at GSI to discuss, among other things, the acceleration of nuclei at the ISR. After all, just a few months earlier, the ISR had accelerated alpha particles (as the high-energy crowd would say) or helium nuclei (as the nuclear physicists would prefer). So why not climb up the nuclear chart and accelerate "real" stuff to "real" energies?

In a summary talk of this workshop, Hans explained how he came to be interested in this field: "Standing up here I wonder how I ever got there. In case somebody has not realized that, I am strictly speaking a low energy nuclear physicist whose horizon, by my own experience, have never yet exceeded 12 MeV/u. Speaking about something one does not understand is already painful, but summarizing something one does not understand is really disastrous. You must ask R. Stock what trick he played to get me on this panel. [...] Suggesting that the ISR could be used for heavy ions is far from being original. People have had such ideas since 1973/1974. [...] My own interest in this, slowly growing during 1980, was continuously discouraged by many arguments, among them that particle physicists just would not at all be interested in it. I obviously didn't talk to the right people. It was my greatest experience during this meeting that [...] particle physicists declared their decided interest. [...] H.G. Fischer in his talk this morning even pleaded in public 'Why don't you join us?.' To this I want to respond as a nuclear physicist."

And respond he did! With his typical pace and dedication, variously described as "breath-taking" (by outside observers) or "breakneck" (by participating students). Special seminars and working sessions were organized with his Heidelberg group to educate them in alien concepts like "rapidity" and "quark soup." Not everybody was immediately enthusiastic to leave known territory and start from scratch, but Hans's enthusiasm was convincing, one way or another. The ones who had no choice had no choice, while others were seduced by various means. One of his students, freshly graduated, was about to leave, horrified by the idea of experiments involving more people than he had fingers on one hand. So Hans persuaded him to stay for a short "test" period at CERN, before helping him to find a new position in nuclear physics. The Swiss Alps and the French wine then did the rest.

Hans's sabbatical at CERN in 1983 coincided not only with the AFS soft lepton venture, but also with the design of the fixed target heavy-ion program at the CERN SPS in general and the HELIOS experiment in particular, of which he became the first spokesperson. The ISR idea was never brought to fruition, nor was a proposal to use the PS, but in 1986, ultra-relativistic heavy-ion physics finally got off to a start on both sides of the Atlantic, with the CERN SPS and the BNL AGS machines. And lepton pairs, both soft and hard, are still (or again) at the forefront of exciting physics. (And incidentally, the AFS measurement did not resolve the soft lepton mystery in pp collisions; it rather added to it.) (Figs. 4.10 and 4.11).

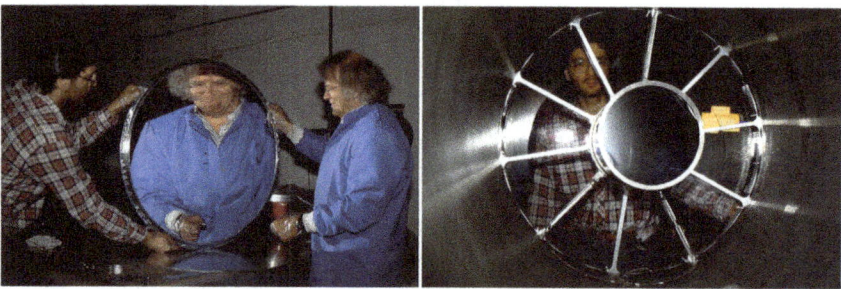

Fig. 4.10 Transitioning to high energy heavy-ion physics at CERN. The CERES proposal 1988. Collection of photos 1990. © Hans J. Specht or Jurgen Schukraft. All rights reserved

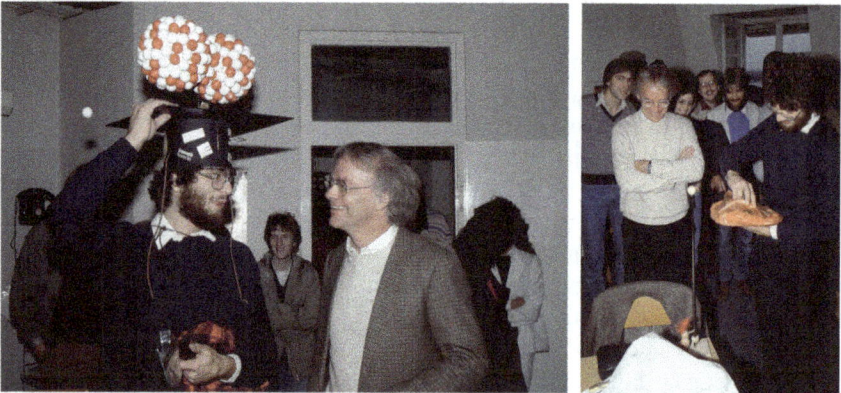

Fig. 4.11 Jurgen Schukraft Ph.D. Thesis in Heidelberg in October 1983 with Hans J. Specht as supervisor. © Hans J. Specht or Jurgen Schukraft. All rights reserved

The Axial Field Spectrometer Years

Michael Albrow

In 1983 Hans Specht, then at the University of Heidelberg, came to CERN for a one-year sabbatical and was invited by Bill Willis to join the Axial Field Spectrometer (AFS) collaboration (experiments R807/R808) at the CERN Intersecting Storage Rings (ISR). This was the first hadron collider, colliding protons on protons with beam energies up to 31.5 GeV, a forerunner of today's Large Hadron Collider whose proton beams have 225 times higher energy. From its inception in 1971 with many "simple" experiments, some using only scintillation counters or lead glass blocks and others using spectrometers with tiny solid angles, we eventually learned how to build full azimuth large solid angle detectors to measure high-energy electrons, photons, muons, and hadron jets from quark and gluon scattering. Furthermore, the versatility of its two independent crossing rings also allowed collisions of protons on antiprotons and on light ions, namely deuterons and alpha particles, as well as α + α collisions at 126 GeV center-of-mass energy.

The α + α collisions happened just before Hans joined the AFS, and the analysis of the data was underway. They were interesting not only for the very higher multiplicity "central" events (how many "wounded nucleons"—any signs of a quark–gluon plasma?), but also for the elastic and quasi-elastic "peripheral" collisions. We had installed some small drift chambers above and below the downstream beam pipes to detect α-particles coherently scattered to small angles. With an elastic-scattering trigger, the display of drift times inspected on shift one evening showed an unexpected dip. It was a diffraction pattern (α-particles also behave as waves!) with the dip at smaller momentum transfer than in proton–proton scattering, the α-particles being bigger than protons. It is not often in particle physics that one can discover a new phenomenon on shift! We also found events where both α-particles were coherently scattered, losing a little energy which appeared as a $\pi + \pi$-pair in the central detectors—a pure double pomeron exchange, still poorly understood. (At the LHC with heavy-ion collisions, double photon exchange ($\gamma + \gamma$) dominates for very small scattering angles.)

It was the combination of these nuclear collisions at unprecedented energies, with the ability to use photons and electrons as probes of new phenomena, that led Hans to join the AFS, even though data-taking was coming to an end. Low energy electron and photon production was showing anomalies, and he was instrumental in adding some small gas Cherenkov counters to distinguish electrons from charged pions. The ISR ceased operation in early 1984 as CERN was gearing up to build the e + e-collider LEP, the SPS had been successfully run as a proton–antiproton collider, and the experiments UA1 and UA2 had found the W and Z bosons. In 1984 I joined UA1, the ultra-relativistic heavy-ion program was approved at the SPS, and Hans was appointed spokesperson of HELIOS/NA34-2. So, my interactions with Hans were not for building detectors or running shifts on the AFS, but for data analysis and paper-writing. When he was at CERN we would share very enjoyable get-togethers to discuss not only the results we were extracting from the ISR data, but also his ideas for the future, especially about collisions of heavier nuclei which are now an important part of the LHC "heavy-ion" program. (This owed much also to Bill Willis, who had been the "godfather" of the AFS and had brought Hans into the ISR experiments.)

Thinking back, had CERN developed the ability to make beams of fully ionized heavy nuclei before the ISR was decommissioned it could perhaps have had a very active second wind, but as things were, the Relativistic Heavy Ion Collider (RHIC) at Brookhaven would take on that field. As the ISR was closed and the LHC was still far in the future, Hans continued his studies with fixed targets of nuclei in the HELIOS experiment at CERN's SPS, becoming spokesperson. This reused some of the AFS detectors, including uranium-scintillator calorimeter modules. With the best ever hadron energy resolution, these had co-discovered (with UA2) high transverse energy jets. At HELIOS/NA34 Hans continued using both soft and hard photons as a probe of new physics, well in advance of the discoveries of quark–gluon plasma as a new state of matter at RHIC and the LHC.

I also remember Hans's enthusiasm as he told me about his ideas for the use of nuclear beams for cancer therapy. I moved away from CERN in 1989, after which we rarely met, but I have fond memories of our conversations during the preceding 6 years.

Recollections of Hans Specht: CERES or the Perpetual Race Against Time

Axel Drees

Hans Specht was a charismatic and brilliant experimentalist with a keen eye for cutting-edge detector concepts and how to apply them most effectively. He had a phenomenal intuition and was relentless in his commitment to the pursuit of science. I was lucky enough to have had him as my mentor and unwavering supporter throughout my career.

Hans and I worked together in Heidelberg for 12 years from early 1986 to the end of 1997—from my doctorate to my habilitation. Those years saw the construction, operation, and upgrade of the CERES experiment at CERN, arguably Hans' experimental masterpiece. The result was the discovery of what is now known to be thermal radiation from the quark–gluon plasma. The great scientific success of CERES became the foundation of my own career, and I am grateful for all I learned from Hans. Remembering him today feels bittersweet.

The following is a collection of memories from my time with Hans. Some of them were adapted for this collection from an earlier text I had written for Hans on the occasion of his 60th birthday.

4 Reflections and Tributes

My First Encounter with Hans Specht

It was September of 1985 in Strasbourg at a conference on "The Quark Structure of Matter." The morning session was on future heavy-ion experiments at the CERN SPS; it promised to be less interesting since nuclear physics was the last thing on my mind. I had settled into the comfortable armchairs in the auditorium of the European Parliament where the conference was hosted. But soon I found myself mesmerized at the edge of my seat listening to a charismatic speaker bursting out at the seams with energy and enthusiasm about the new program and the HELIOS experiment.

Just moments earlier my trajectory had been crystal clear. I would graduate in a few months with a diploma thesis in experimental high-energy physics from the University of Bonn, and then I would join the group of Walter Blume at the MPI in Munich and pursue a doctorate with the ALEPH experiment. The offer was all but signed.

By the end of the talk I had made up my mind. Change of plan: I was going to go to Heidelberg to learn more from Hans Specht.

Working with Hans Specht

Hans trusted his intuition about people and did not hesitate to entrust junior members of his group with challenging projects that needed to be completed on a rigorous schedule. Of the many examples, I will just mention my own initial experience here: with little prior knowledge I was assigned the task of setting up the data acquisition for the test beam at CERN, due only a few months after I joined the group. His trust combined with high expectations empowered my confidence and that of others. It challenged us to do our best. Going hand in hand with this was the way Hans regarded everyone in the group as fellow scientists, albeit some less experienced than others. For me that created an atmosphere in which I felt my voice was heard, and my contributions counted. I genuinely felt we were working with him as a team. And soon enough most of us would find ourselves following him on a path to achieve what most others deemed impossible. That mindset, paired with our youthful enthusiasm, was the key pillar in our success: CERES could not have been built without it.

Nonetheless, it was often difficult to live up to Hans' standards and expectations, in part because he set the bar so high for himself. After all, as he himself would point out to us, he was the only one in the group that worked 12 h a day 7 days a week. He definitely was leading by example, but except for critical periods of the experiment, by and large most of us were simply unable to sustain such a schedule for extended periods of time. And occasionally he needed to hear that. One such occasion was when he became Director of GSI. Naturally, he was now busier than ever, yet he wanted to remain in touch with the progress of CERES and the group. He proposed what seemed an ideal solution to him: moving forward we would have our regular weekly group meeting on Saturday mornings. The rest of the group was less enthusiastic about spending part of their Saturdays that way for the next 5 years. It fell on me to have the difficult conversation with him that we were not going to cross this line. As always in these moments, Hans accepted the facts, and we moved on.

Of the many things I admire about Hans, I want to highlight his phenomenal intuition and his passionate work with paper and pencil. You would rarely find him without a pencil ready to go in his pocket, usually already worked down to a stub. Most of the design of CERES was done by Hans on hundreds of pages of meticulously organized handwritten notes including graphs and tables, which he would happily share with us. Unlike today, Monte Carlo computer simulations played a minor role in the development of CERES. While he would also seek the support of computer simulations, always carried out by others, he approached them with significant skepticism. One of his mantras was that if the computational results did not agree with his hand calculations, for sure the computer-generated results were wrong. And indeed, it was rare that Hans had overlooked something in his estimates that would then be brought to light by a computational result. I did not adopt the habit of carrying pencils around in my pockets, but the rest of his approach I enthusiastically embraced and still practice today. From time to time, I am successful in passing what I learned from Hans on to the next generation (Fig. 4.12).

As the years went by, I took more and more of a leading role in the CERES group. This was not spelled out initially but seemed to happen naturally. Once Hans took on responsibilities as director at GSI it became more formalized, and I effectively led the daily operation of the group as it moved forward. Growing into that role did not come without worries for me, and at times I was riddled by doubts about my ability to succeed. It was in those moments that Hans' steadfast support was the most impactful. I vividly remember one conversation we had; it must have been in the early 1990s. I was ready to give up academia. Hans listened to what I had to

Fig. 4.12 From left to right: Axel Drees, Hans-Werner Bartels, and Hans J. Specht in June 1989, shortly after Axel's Ph.D. defense. © Axel Drees. All rights reserved

say and then diffused my doubts by calmly declaring "Axel wir alle kochen nur mit Wasser"—Axel we all cook with water.

Simply Put: More Luck than Sense

Sometimes talent and work ethics are insufficient to succeed, and we need just a bit of good fortune. The following incident is one such occasion which could easily have cut CERES short, were it not for pure luck. This incident is a prime example of a chain of errors so sophisticated and insidious that it is hard to imagine. The story began like this: the high voltage of the UV detectors had been turned off for the night since there was no beam. And because there was no beam, we had decided to cancel the night shift. As it turned out, it was a big mistake not to have someone in the control room to intervene in the sequence of hardware and software failures that would unfold.

In the middle of the night, due to some hardware glitch, the high-voltage power supply issued a false alarm, which would usually indicate a discharge in the UV detector. But because the high voltage was off, there could not have been a discharge. Such an event was not foreseen, and the high-voltage control software sprang into action and initiated the usual autonomous recovery procedure. The power supply was reset, and the high voltage was brought back to operating conditions. Since the detectors were under normal gas flow, this could have been all right. However, due to unlikely circumstances involving both software and hardware, some high-voltage channels remained at zero. The result was a continuous discharge that burned for many hours until the first physicists showed up in the morning and rubbed their eyes in astonishment.

This catastrophe should have completely destroyed the detector, but miraculously the detector was fine. Nonetheless, this mishap confirmed Hans' profound distrust towards anything that came out of a computer. He was beside himself and personally conducted a high-stress interrogation of the responsible parties, who confessed but pleaded that events were beyond their control.

CERES or the Perpetual Race Against Time

When I joined Hans' group, he had already started planning an electron pair spectrometer extension for the HELIOS detector. This nameless upgrade project would in due time become the widely discussed experiment known as CERES. The key component of the upgrade was a ring-imaging Cherenkov counter (RICH) combined with a gaseous UV light detector. After a successful test beam with a RICH prototype in the summer of 1986, Hans established a bold 2-year timeline for the HELIOS upgrade:

- March 1987: Finalize the pad readout for the UV detectors
- June–August 1987: Test a full-scale prototype at CERN
- February 1988: Complete the construction of the final UV detectors
- March–August 1988: Installation and tests within HELIOS
- September 1988: Take first physics data

The tight planning epitomized Hans' impatience and optimism. It gave a sneak preview of how he would pursue CERES in the following years. The expectations, especially regarding the pad readout for the UV detectors, were particularly ambitious since Hans had proposed the concept only a few weeks earlier. It replaced the projective wire readout that had proven unsuitable for ring imaging in the test beam. With unshakable trust in the talent of his students, he assigned the development of the pad readout as a diploma thesis project. At the time it seemed unsurprising to me that only a few months later, in the summer of '87, the new pad readout was tested with a beam at CERN. The results were very promising. But despite the success, most of the HELIOS collaboration remained utterly unimpressed and in no way shared Hans' confidence. As a result, their paths diverged, and CERES was born.

The electron spectrometer, no longer limited in space by the need to fit into the existing HELIOS apparatus, was optimized as a standalone dilepton experiment—CERES. The original CERES design could not have been more minimalistic and ambitious. It relied almost exclusively on the RICH detector technology, which so far had only been deployed with limited success in other experiments. In CERES, two RICH detectors would identify and track electrons: one RICH before and one after a magnetic field with a unique configuration that preserved the ring images. The only other detector was a silicon pad detector that served as an interaction trigger.

The CERES proposal was formally submitted to CERN on 18 June 1988, as experimental proposal P237, with a small yet enthusiastic collaboration of no more than 22 members from the University of Heidelberg, the Max Planck Institute in Heidelberg, and the Weizmann Institute in Israel. In the spring of 1990, space at the H8 beamline, just upstream of HELIOS, was cleared to make room for NA45, the officially approved code for CERES. Many components had already been built and no time was wasted in starting construction of the apparatus. Naturally, Hans wanted to take the first physics data that same autumn, so it was not surprising that his impatience grew as the beam time approached. It was a race against time. As miracles are known not to happen by themselves, we all worked harder than ever. After all, we were in the critical phase of the experiment. Little did we know then how many critical phases this experiment had in store for us.

Initially, our efforts were rewarded, and we celebrated the first Cherenkov rings from S–Au collisions in 1990. Unlike previous RICH detectors, the number of photons and the resolution of the CERES rings met expectations. But not everything went as smoothly as hoped. Exposed to the beam, the UV detectors were sparking almost continuously. This limited the event rate to less than one Hz—a couple of events per spill. But even worse, the sparking was destroying the readout electronics, reducing its lifespan to a few days, a major disaster. Additionally, the recently published lepton pair measurements from HELIOS gave little cause for joy, showing that in p-Be collisions, all lepton pairs could be explained by hadron decays within systematic errors. This ended a more than 15-year hunt for anomalous lepton pairs and deprived CERES of a *raison d'être*.

The year 1991 was mainly spent studying the causes of the excessive sparking of the UV detectors and figuring out how to eliminate it. The central issue was the parallel-plate amplification scheme used in the UV detectors, which apparently could not tolerate the high radiation background from heavy-ion beams. Attempts to mitigate the parallel-plate amplification with resistive cathode planes failed. Only returning to a wire amplification scheme could bring about the long-awaited breakthrough. We published a paper documenting our ordeal and Hans insisted on concluding the paper with the strong statement: "We will in future stick to the lovely tolerance of a wire amplification scheme."

Solving the spark breakdown issue had another welcome side effect. The UV detectors no longer needed to be operated in a gated mode, which required switching the high voltage from standby to operating condition many times per second. This switching or pulsing of the high voltage had proven a perennial source of trouble. At times Hans would become so frustrated with the performance of the high-voltage system that he proposed to ban the responsible colleagues from the control room, or offer them free rides to the moon.

Finally, in 1992, the first (and last) beam time with sulfur ions arrived. Data was coming in, and we were looking forward to the completion of a successful running campaign. But the next race against time was just lurking around the corner. Data taking would resume in only a few months, and Hans had grand new plans for the forthcoming running campaign. He wanted to improve HELIOS' measurement of e^+e^- pairs from p-Be collisions. That could not be done without help, so he convinced Volker Metag, another of his mentees, to bring his electromagnetic calorimeter, TAPS, from GSI and put it behind the CERES spectrometer. The time window was very short to set up TAPS at CERN and integrate it into our data acquisition system. In addition, we also needed an e^+e^- pair trigger system to record sufficiently large data samples with p-Be collisions. This so-called intermediate level trigger would become the final component that Hans had foreseen for CERES, even if not as originally planned. Naturally, one of Hans' students was tasked with its realization.

Data from p-Be collisions was streaming in, and we were extracting initial results in real time. It was another roaring success! As usual, time to relish our accomplishments was fleeting and reality hit us soon enough in the form of the next critical phase of the experiment. By the end of the year, a proposal for the continuation of CERES with a lead beam had to be submitted. Although the team was completely exhausted, Hans somehow managed to get us all on board for this renewed Herculean effort. Over the holidays 1993–94, a comprehensive review of the current CERES results and a new proposal (P280) were prepared, and then submitted to CERN on 22 January 1994. The main proposed improvement of the CERES setup was the addition of external tracking to aid the ring recognition in the lead beam environment. Of course, Hans chose a minimalistic approach to the problem: two space points in front and one behind the spectrometer. Furthermore, the two space points in front would be provided by radial silicon drift detectors, an *avant-garde* detector technology, not yet tested in a physics experiment.

In the meantime, the S–Au data analysis proved to be as challenging as getting the original detector to work in the first place! But by the beginning of 1994, it was also largely completed. CERES had discovered a significant enhancement of e^+e^- pairs from S–Au collisions than a simple extrapolation from p-Be data would suggest. This result triggered considerable interest and numerous theoretical studies. As we now know, it was the first experimental observation of thermal radiation from the quark–gluon plasma. It inspired the NA60 experiment and new follow-up experiments are still being planned today, 30 years later. The results were published in Physics Review Letters in 1995. By now the paper has been cited nearly 1000 times, making it the most highly cited experimental paper from the SPS heavy-ion program.

The year 1994 was a turning point for CERES. Not only did we present the first high-impact physics results from S–Au, but somehow, we also managed to complete the upgrade of the experiment. The SPSLC, the CERN committee responsible for allocating beam time, remained skeptical. Initially, they had granted us only nine days of lead beam time at the end of the year. We were under pressure to demonstrate the feasibility of the experiment, since the SPSLC had promised "to have a hard look at the results" before allocating any more beam time to CERES. The results we achieved in nine days exceeded even our own expectations and evidently convinced the doubters on the SPSLC committee.

CERES collected physics data sets for Pb-Au collisions over two consecutive years. After the first data collection, at the end of 1995, all effort was directed towards data analysis. The goal was to show the first results at the next Quark Matter meeting, the most important conference in the field. Given that it would be held on our home turf in Heidelberg, it was unthinkable that we would not be ready. The situation was complicated by the fact that we had decided to revamp the analysis code for the lead beam era, including a full conversion of the code from FORTRAN to C++. In the five months between the end of data collection and the start of the conference, the software was converted, all raw data—back then a huge volume of about a terabyte—was processed and evaluated. We made it with only five days to spare before the beginning of the conference. We had overcome yet another critical phase of the experiment. It was to be my last before I started my own research group at Stony Brook University.

Fig. 4.13 Raising a glass in celebration of Hans J. Specht's 80th birthday at the Festkolloquium in Heidelberg, June 2016. Left panel: Axel Drees; Middle panel: From right to left—Jurgen Schukraft, Axel Drees, and Thomas Ullrich; Right panel: Hans J. Specht at the Festkolloquium (with J. Schukraft and D. von Harrach as keynote speakers). © Axel Drees and Hans J. Specht. All rights reserved

My Last Encounter with Hans Specht

After I moved on from Heidelberg, our paths crossed regularly over the years (Fig. 4.13). I believe the last time we met in person was in November 2018 at a workshop at the ECT* in Trento on "Electromagnetic Radiation from Dense Hadronic Matter," a topic that remained dear to his heart for nearly 40 years. Though he had visibly aged and was less agile physically, his charismatic style and laser-sharp focus on the pursuit of science were as captivating as ever.

I still chuckle about the surprise email I received from the organizers apologizing that they had booked me in the wrong hotel. They had only recently learned that I was the designated driver of Hans Specht, and they had to change my reservation so that we would be at the same hotel. It was welcome news to meet Hans again, but it was surely new news to me. Well, I guess once you are a student of Hans Specht's you will always remain his student.

Dilepton Radiation from Heavy-Ion Collisions—A Tribute to Hans Specht

Jochen Wambach

1. Introduction

The exploration of the spectral properties of hadrons in extreme conditions of temperature and density has been central to strong-interaction physics for decades. It relates to basic properties of QCD such as spontaneously broken chiral symmetry in the light quark sector and its restoration at high temperature and density, as well as the fate of confinement and the ultimate liberation of quarks and gluons.

Experimentally, extreme QCD matter is studied in heavy-ion collisions at ultra-relativistic energies by measuring and analyzing a variety of observables. Among them, real and virtual photons, emitted from the reaction zone, are particularly precise tools since the photon mean-free path is orders of magnitude larger than the size of the fireball. Thus, they carry undisturbed information about the space–time evolution of the entire collision from the pre-equilibrium stages through the hydrodynamic phases of strongly coupled quarks and gluons and their transition to confined hadrons to the final free streaming.

Historically, two central questions have been in the focus of studies with electromagnetic (EM) probes:

- What are the signatures of chiral symmetry restoration of the observed dilepton spectra?
- Can the formation of the quark–gluon plasma be detected through EM radiation?

As I will discuss, the groundbreaking experiments of the NA60 collaboration at the CERN SPS under the leadership of Hans Specht combined with advanced theoretical calculations have made decisive contributions in our understanding of the properties of QCD matter.

2. Photons from a Thermal Medium

To set the stage, let me start with some general considerations on the in-medium propagation of photons. In the following it will be assumed that the strong-interaction matter is in (local) thermodynamic equilibrium, which pertains to the hydrodynamic phases of the space–time evolution of the fireball. For simplicity (relaxed in actual comparisons of theory with experiment), consider a static isotropic QCD medium at temperature T and baryon chemical potential μ_B. The relevant quantity describing the propagation of a real or virtual photon with 4-momentum $q^\mu = (q_0, \vec{q})$ is the photon polarization tensor $\Pi_\gamma^{\mu\nu}$, which is related to the (retarded) EM current–current correlation function

$$\Pi_\gamma^{\mu\nu}(q_0, |\vec{q}|; T, \mu_B) = -i \int d^4 x\, e^{-iqx} \Theta(x_0) \langle\langle [j_{\text{EM}}^\mu(x), j_{\text{EM}}^\nu(0)] \rangle\rangle, \quad (4.1)$$

where $\langle\langle . \rangle\rangle$ denotes the thermal average in the grand canonical ensemble. Since Lorentz invariance is broken by the medium, there is a separate dependence on q_0 and the modulus of the 3-momentum. The photon spectral tensor, $\varrho_\gamma^{\mu\nu} = -2\,\text{Im}\,\Pi_\gamma^{\mu\nu}$, determines the measurable dilepton emission rate as

$$\frac{dN_{ll}}{d^4 x\, d^4 q} = \frac{\alpha^2}{6\pi^3 M^2} \left(1 + \frac{2 m_l^2}{M^2}\right) \sqrt{1 - \frac{4 m_l^2}{M^2}}\, f^B(q_0; T)\, g_{\mu\nu} \varrho_\gamma^{\mu\nu}(M, |\vec{q}|; T, \mu_B), \quad (4.2)$$

where $M = \sqrt{q_0^2 - \vec{q}^{\,2}}$ denotes the invariant mass of a dilepton pair, m_l the lepton mass, $f^B(q_0; T) = 1/(e^{q_0/T} - 1)$ the thermal Bose function, and $\alpha = e^2/4\pi$ the fine-structure constant.

Using the four-dimensional projectors for a spin-1 particle, $P_{L,T}^{\mu\nu}$, one can further decompose $\varrho_\gamma^{\mu\nu}$ into its longitudinal and transverse components as [1, 2]

$$\varrho_\gamma^{\mu\nu} = \varrho_L P_L^{\mu\nu} + \varrho_T P_T^{\mu\nu}, \quad (4.3)$$

yielding $g_{\mu\nu} \varrho_\gamma^{\mu\nu} = \varrho_L + 2\varrho_T$. At vanishing 3-momentum relative to the heat bath, one has $\varrho_T = \varrho_L$ for all M. At finite $|\vec{q}|$, this no longer holds. For real photons, ($M = 0$), ϱ_L vanishes and the emission rate is purely transverse. An important case is the time-like limit $q_0 \to 0$, $|\vec{q}| = 0$, which determines the electrical conductivity of the medium as

$$\sigma_{el}(T, \mu_B) = \frac{e^2}{6} \lim_{q_0 \to 0} g_{\mu\nu} \varrho_\gamma^{\mu\nu}(q_0, |\vec{q}| = 0)/q_0 = \frac{e^2}{2} \lim_{q_0 \to 0} \varrho_T(q_0, |\vec{q}| = 0)/q_0, \quad (4.4)$$

since at $|\vec{q}| = 0$, $\varrho_L = \varrho_T$. This is a fundamental transport coefficient of QCD matter.

The dependency of the dilepton production rate [3–5]

$$\frac{dN_{ll}}{d^4x\,d^4q\,d\Omega_l} = \frac{\alpha^2}{32\pi^4}\frac{1}{M^4}\sqrt{1-\frac{4m_l^2}{M^2}}f^B L_{\mu\nu}\varrho_\gamma^{\mu\nu} \qquad (4.5)$$

on the lepton angle $\Omega_l = (\phi_l, \theta_l)$ in the photon rest frame yields additional information about the physical processes involved in the thermal emission. With the lepton tensor

$$L^{\mu\nu} = 2\big(q^2 g^{\mu\nu} - q^\mu q^\nu + \Delta l^\mu \Delta l^\nu\big), \qquad (4.6)$$

where $\Delta l^\mu = l^{+\mu} - l^{-\mu}$, and $l^\pm$ are the lepton four-momenta, the angular distribution takes the form

$$\frac{dN_{ll}}{d^4x\,d^4q\,d\Omega_l} \propto \frac{1}{3+\lambda_\theta}(1+\lambda_\theta\cos^2\theta_l$$
$$+\lambda_\phi\sin^2\theta_l\cos 2\phi_l + \lambda_{\theta\phi}\sin 2\theta_l\cos\phi_l$$
$$+\lambda_\phi^\perp\sin^2\theta_l\sin 2\phi_l + \lambda_{\theta\phi}^\perp\sin 2\theta_l\sin\phi_l) \qquad (4.7)$$

where the λ's are the anisotropy coefficients.

For a static medium, the only non-vanishing coefficient is λ_θ, which can be expressed as

$$\lambda_\theta(M, |\vec{q}|) = \frac{\varrho_T - \varrho_L}{\varrho_T + \varrho_L} \qquad (4.8)$$

for $m_l \ll M$.

3. Hadronic Many-Body Theory

In the following, I discuss the evaluation of the emission rates and polarization observables in a theory for hadronic matter developed more than 20 years ago [6]. Over the years, this hadronic many-body theory (HMT) has proven to account well for a large number dilepton experiments and, in particular, the measurements of the NA60 collaboration. It is fair to say that the latter have played a significant role in the early acceptance of the theory and its underlying physical ingredients.

In the kinematic range of the NA60 experiments, the EM current is carried by the lightest quarks, u, d, s:

$$j_{EM}^\mu = \frac{2}{3}\bar{u}\gamma^\mu u - \frac{1}{3}\bar{d}\gamma^\mu d - \frac{1}{3}\bar{s}\gamma^\mu s \qquad (4.9)$$

which can be rearranged into hadronic states as

$$j_{\text{EM}}^\mu = \frac{1}{\sqrt{2}} j_\rho^\mu - \frac{1}{3\sqrt{2}} j_\omega^\mu - \frac{1}{3} j_\phi^\mu \tag{4.10}$$

with properly normalized hadronic vector currents $j_V^\mu (V = \rho, \omega, \phi)$ with isospin $I = 1$ (ρ) and $I = 0$ (ω and ϕ). Since the currents enter the polarization tensor $\Pi_\gamma^{\mu\nu}$ quadratically, the relative weights for ($\rho : \omega : \phi$) are (1 : 1/9 : 2/9). Using vector dominance [7], the EM current at invariant masses $M \leq 1$ can thus be expressed in terms of light vector mesons, giving rise to the current field identity

$$j_{\text{EM}}^\mu (M \leq 1 \, \text{GeV}) = \frac{m_\rho^2}{g_\rho} \rho^\mu + \frac{m_\omega^2}{g_\omega} \omega^\mu + \frac{m_\phi^2}{g_\phi} \phi^\mu \tag{4.11}$$

with the vector-meson fields ρ^μ, ω^μ, and ϕ^μ as the relevant degrees of freedom.

Given the relative weights of the vector-meson channels, j_{EM}^μ is saturated by the $I = 1$ ρ-meson to a very good approximation. The photon field A^μ then couples exclusively to the isopin 3-component of the ρ-field

$$\mathcal{L}_{\text{int}}^{\text{VDM}} = \frac{e m_\rho^2}{g_\rho} \rho_3^\mu A_\mu. \tag{4.12}$$

Hence the propagation of the ρ-meson through the QCD medium is of central interest.

Since the ρ-meson predominately decays into two pions, the starting point for building a theory of the ρ-meson propagator is an effective $\pi\rho$ Lagrangian for which the parameters m_ρ, g_ρ and a hadronic form factor cutoff Λ_ρ are fitted to the experimental p-wave $\pi\pi$ phase shifts, the pion-EM form factor, the $e^+e^- \to$ hadrons cross sections, and data from hadronic τ decays. An obvious medium modification of the ρ-meson stems from the modification of its pion cloud through interactions with the surrounding hadronic matter, resulting from a pion propagator modified by couplings with baryons and mesons. The resummed in-medium pion propagator was much discussed in the 1980s in connection with the possibility of p-wave pion condensation in nuclear and neutron matter [8]. A variety of experiments on heavy nuclei have established that this does not occur at nuclear saturation density, due to the short-range repulsive parts of the nucleon–nucleon interaction, often parametrized by the 'Migdal parameter' g'. In addition, one has to include direct couplings of the ρ-meson with baryonic resonances B^* such as $N^*(1520)$, $N^*(1720)$, and so on. Their effective Lagrangians are constrained by empirical decay branchings and scattering data as well as photo-absorption on the nucleon and nuclei. For the latter, it was shown that the direct $N\rho - B^*$ couplings provide the resonant contributions, while the pion cloud yields the non-resonant part of the cross section.

Cutting the emerging diagrams of the in-medium ρ-meson self-energy, one can identify a variety of elementary in-medium processes, such as resonance decays, meson-exchange scattering, nucleon–nucleon bremsstrahlung, etc. They provide the dominant contributions to the spectral functions for invariant masses below the $\pi\pi$-threshold, and thus represent the low-energy and low-momentum excitations of the medium. As such, they are crucial for the electrical conductivity σ_{el} [9].

At first sight, it is not obvious how the in-medium restoration of spontaneously broken chiral symmetry manifests itself in a purely hadronic description [10]. Upon closer inspection, it becomes clear, however, that processes involving the capture of pions from the heat bath, such as $\pi\rho \to a_1$, lead to a "mixing" of hadronic parity partners. When chiral symmetry is restored, this mixing leads to a degeneracy in which the relevant spectral functions become identical for all photon momenta $q^\mu = (q_0, \vec{q}\,)$. Exact degeneracy only holds in the chiral limit of vanishing quark masses.

4. Extracting Thermal Dilepton Radiation from Heavy-Ion Data

An experimental milestone was reached in 2006 and 2008 by the NA60 collaboration dimuon measurements in 158 AGeV ($\sqrt{s_{NN}} = 17.3$ GeV) In-In collisions at the CERN SPS [11, 12]. The high statistics of the experiments and a good mass resolution of about 2% allowed for an unambiguous extraction of the excess radiation in the collision for invariant masses $M < 1.4$ GeV from the total inclusive dimuon yields (Fig. 4.14). It should be noted that there is no yield below the dimuon threshold $2m_\mu \sim 211$ MeV because of the lepton phase space factor in Eq. 4.2. Therefore it is not possible to extrapolate dimuon data to the time-like limit $q_0 \to 0, |\vec{q}\,| = 0$, which determines the electrical conductivity. This can only be achieved with dielectrons, where $2m_e \sim 1$ MeV.

The data revealed substantial broadening of the in-medium ρ-meson without a significant change in the centroid of the mass distribution. These observations laid to rest suggestions that the ρ-meson, and for that matter all light hadrons, lose a large fraction of their mass during the chiral restoration transition [13–15].

The measurements were later extended to invariant masses $M < 2.6$ GeV in [16, 17], where the M-dependence of the inverse slope parameter T_{eff} of the excess dimuon spectra was extracted (Fig. 4.15). Since $T_{\text{eff}} \approx T + Mv^2$, where T is the true temperature and v the average radial flow velocity of the fireball, the linear rise up to about 1 GeV is consistent with the expected flow of hadrons. For $M > 1$ GeV, the values rapidly drop by about 50 MeV and remain roughly constant at around 200 MeV, well above the pseudo-critical temperature for chiral symmetry restoration, T_χ^{pc}, which was later determined from LQCD to be ~ 156 MeV [18] (and references therein). The near constancy of T_{eff} implies the absence of flow effects and the radiation must therefore originate from the early stages of the collisions. Thus it essentially measures the true temperature of the initial phase. In addition, the thermal yields are compatible with those expected from (nearly) perturbative u, d, s quark sources (Eq. 4.9). These conclusions emerged from intense discussions led by Hans Specht (some of which I vividly remember).

Another groundbreaking result from the NA60 collaboration was the measurement of the angular distributions of excess dimuons in two invariant-mass bins, requiring very high statistics [19]. The results are displayed in Fig. 4.16. The near absence of any polarization was attributed to a complete randomization of the thermal system. This is probably not correct, as recent results from HMT show (see below).

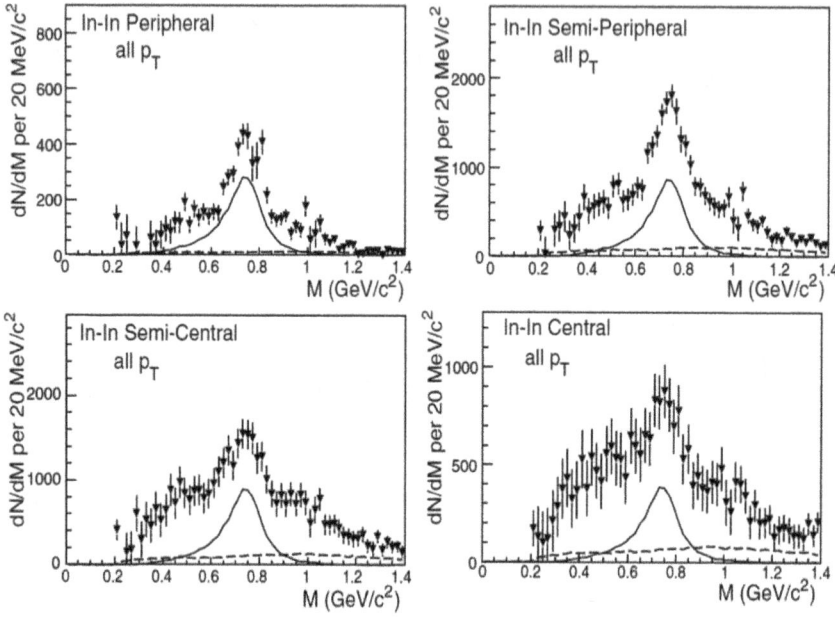

Fig. 4.14 Centrality dependence of the excess dimuon mass spectra from [11]. For comparison the contributions from the cocktail ρ and uncorrelated charm decays are shown. © 2006 American Physical Society. All rights reserved. Phys. Rev. Lett. 96 (2006) 162302

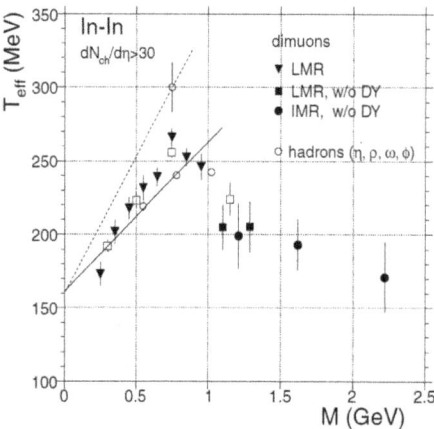

Fig. 4.15 Inverse slope parameters T_{eff} of the excess dimuon mass spectra for $0.2 < M < 2.5$ GeV. Results from hadronic flow measurements are shown for comparison. © 2008 American Physical Society. All rights reserved. Phys. Rev. Lett. 100 (2008) 022302

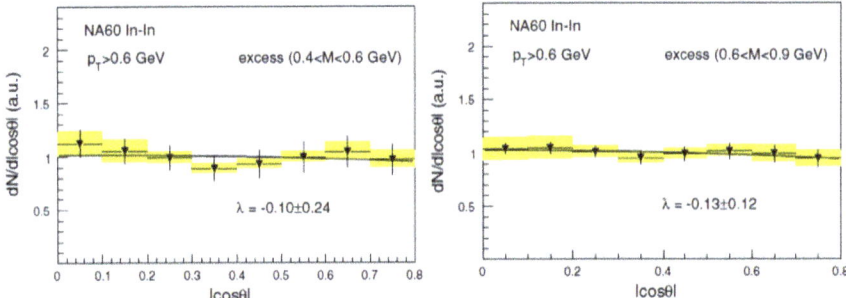

Fig. 4.16 Dimuon angular distributions measured by the NA60 collaboration in the Collins-Soper frame for In-In collisions [19] in two invariant-mass bins. © 2008 American Physical Society. All rights reserved. Phys. Rev. Lett. 102 (2009) 222301

5. Where Do We Stand?

A detailed comparison of the high-statistics dimuon measurements in central In-In collisions by the NA60 collaboration at the CERN SPS [11, 12] with an HMT calculation [21] which includes a large set of hadronic sources, effects of chiral mixing, and LQCD improved $\bar{q}q$ rates [20], shows excellent agreement for the background subtracted rates in the ρ-meson region (left panel of Fig. 4.17) [21]. The total subtracted yields [22] (right panel of Fig. 4.17) based on the same spectral functions and a fireball evolution model calibrating the LQCD equation of state are also in good agreement with the NA60 data. Utilizing the fact that, in the region $M > 1.5$ GeV, the medium effects on the EM spectral functions are small, one concludes that the thermal rates are independent of the collective flow of the fireball. Thus the high-mass region enhances the sensitivity to the early phases of the collision, and a rather precise thermometer could be established [22, 23].

Some general conclusions can be drawn from these precise comparisons:

1. A ρ-meson, propagating in the thermal fireball, suffers collisional broadening which becomes very large near and above T_χ^{pc} and merges smoothly into the $\bar{q}q$ continuum. The centroid of the invariant-mass distribution remains essentially temperature-independent, indicating that the ρ-meson pole mass is barely shifted. In the meantime, this is firmly established theoretically from first-principles LQCD calculations of the temperature dependence of the ρ-meson screening mass [24].

2. The high precision of the NA60 experiments clearly shows the $\rho - a_1$ "chiral mixing" effect above $M > 1$ GeV. The restoration of chiral symmetry in the QCD medium must lead to a (nearly) complete degeneracy of the real-time spectral functions of hadronic parity partners, especially those of the ρ- and a_1-mesons [25]. This was confirmed in a two-flavor quark–meson model [26] which showed a merging of the pole masses after symmetry restoration (Fig. 4.18). In fact, it is found in [26] that the ρ and a_1 spectral functions are degenerate for all energies and 3-momenta, in accordance with general QCD expectations.

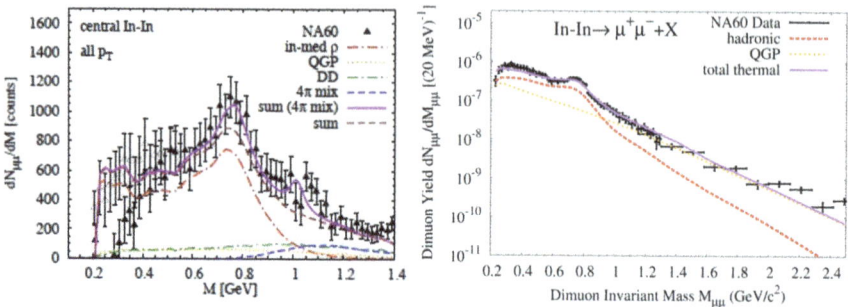

Fig. 4.17 The subtracted excess dimuon spectrum in the invariant-mass region of the ρ-meson and the total excess yield (right panel) as measured by the NA60 Collaboration [11] is compared to theoretical results from Ref. [21] (left panel) and Ref. [22] (right panel). (Left panel) CC-BY-NC (right panel) CC-BY-4.0

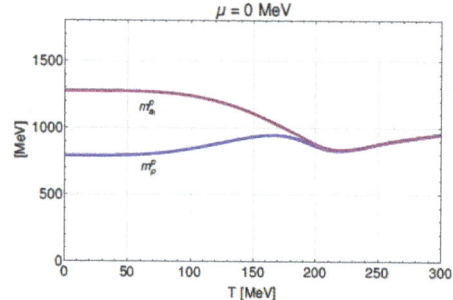

Fig. 4.18 Temperature dependence of the ρ- and a_1 pole masses [26]. @ 2017 American Physical Society. All rights reserved

By now, "parity doubling" in the meson sector is firmly established in LQCD studies of temperature-dependent screening masses of parity partners involving light $\bar{u}d$ quarks [24]. Similarly, parity doubling has been found in lattice studies of the screening masses of baryon octet and decuplet states [27]. The latter results are highly relevant for the high-μ_B region of the QCD phase diagram.

Recently [28], HMT has also been applied to the dimuon angular distribution measurements of the NA60 collaboration [19] (Fig. 4.16).

For the space–time evolution of the collision, an isentropically expanding fireball model has been used, whose particle content reproduces the experimentally observed light-hadron data and whose time evolution was constrained by observed p_T spectra and hydrodynamic expansion time scales. The model includes a QGP phase that reaches maximal initial temperatures close to $T = 240$ MeV. The theoretical predictions (in the Collins–Soper frame) describe the measured angular distributions in θ_l (and ϕ_l) quite well (Fig. 4.19). Note, however, that according to the calculation, the near absence of a net polarization (rather flat angular distributions) is not related to thermal isotropy arguments but stems from the properties of the equilibrium EM spectral functions ϱ_L and ϱ_T, which yield a rather small net polarization in the mass region around $\sim$0.5 GeV for In + In collision energies.

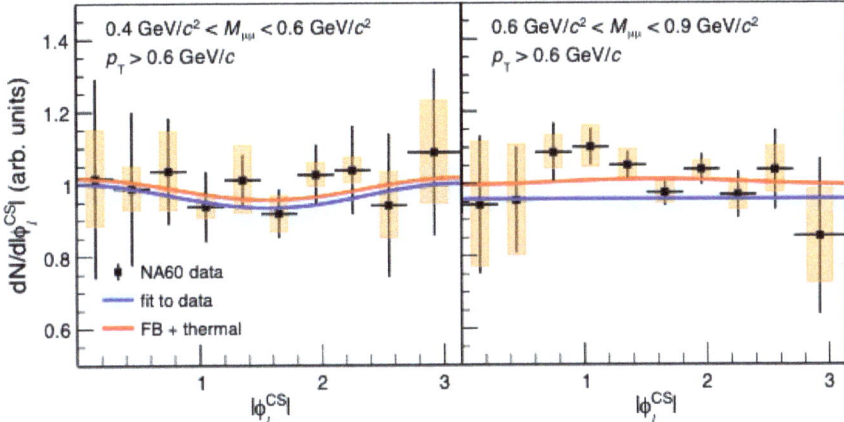

Fig. 4.19 Dimuon angular distributions (red lines) calculated in a fireball (FB) model combined with HMT thermal rates, integrated over two dimuon mass windows, in the Collins-Soper frame. The results are compared to NA60 data [19]. The blue lines represent fits by NA60. CC-BY-4.0

6. Conclusions

Putting the achievements of the NA60 collaboration in perspective, I have tried to convey the enormous impact these high-quality measurements have had and continue to have in sharpening our understanding of:

- the in-medium propagation of the ρ-meson,
- signals of chiral symmetry restoration in thermal dimuon spectra,
- evidence for an equilibrated early partonic phase in relativistic heavy-ion collisions,
- the determination of the initial temperatures of the fireball,
- features of the angular distribution of dimuons in a heavy-ion collisions.

It is fair to say that the huge success of the dilepton programs at CERN, and here in particular the NA60 experimental campaign, was distinctively shaped by Hans Specht and his scientific vision. He was one of the key proponents of the CERN ultra-relativistic heavy-ion program at the SPS accelerator. From many discussions, I came to appreciate his deep understanding of physics, his remarkable familiarity with theoretical concepts, and his determination to achieve scientific goals he believed in. His scientific leadership will be missed by all of us.

Acknowledgments This work was partly supported by the DFG Collaborative Research Centre 315477589-TRR 211, "Strong interaction matter under extreme conditions."

References

1. McLerran L, Toimela T 1985 *Phys. Rev. D* **31** *541*
2. Landsman N P, van Weert C G 1987 it *Phys. Rep.* **145** *141*
3. Bratkovskaya E L, Teryaev O V and Toneev V D 1995 *Phys. Lett. B* **348** 283

4. Baym G, Hatsuda, T Hatsuda and Strickland M 2017 *Phys. Rev. C* **95** *044907*
5. Speranza E, Jaiswal A and Friman B (2018) *Phys. Lett. B* **782** *395*
6. Rapp R and Wambach J 2000 *Adv. Nucl. Phys.* **25** *1*
7. Sakurai J J 1960 *Annals of Physics* **11** *1*
8. Migdal A B 1978 *Rev. Mod. Phys.* **50** *107*
9. Rapp R 2024 [arXiv:2406.14656 [hep-ph]
10. Rapp R and Wambach J 1999 *Eur. Phys. J. A* **6** *415*
11. Arnaldi A, et al. (NA60 Collaboration) 2006 *Phys. Rev. Lett.* **96** *162302*
12. Arnaldi A, et al. (NA60 Collaboration) 2008 *Phys. Rev. Lett.* **100** *022309*
13. Brown G E and Rho M 1991 *Phys. Rev. Lett.* **66** *2720*
14. Hatsuda T and Lee S H 1992 *Phys. Rev. C* **46** *R34*
15. Brown G E and Rho M 2002 *Phys. Rep.* **363** *85*
16. Arnaldi A, et al. (NA60 Collaboration) 2009 *Eur. Phys. J. C* **59** *607*
17. Arnaldi A, et al. (NA60 Collaboration) 2009 *Eur. Phys. J. C* **61** *711*
18. Bazavov A, et al. 2019 *Phys. Lett. B* **795** *15*
19. Arnaldi A, et al. (NA60 Collaboration) 2009 *Phys. Rev. Lett.* **102** *222301*
20. R. Rapp 2013 *Adv. High Energy Phys.* *148253*
21. Rapp R, van Hees H and Strong T 2007 *Braz. J. Phys.* **37** *779*
22. Rapp R, van Hees H 2016 *Phys. Lett. B* **753** *586*
23. The Hades Collaboration 2019 *Nat. Phys.* **15** *1040*
24. Bazavov A., et al. 2019 *Phys. Rev. D* **100** *094510*
25. Hohler P M and Rapp R 2014 *Phys. Lett. B* **731** *103*
26. Jung C, Rennecke F, Tripolt R-A, von Smekal L and Wambach J 2017 *Phys. Rev. D* **95** *036020*
27. Aarts G, et al. 2019 *Phys. Rev. D* **99** *074503*
28. Seck F, et al. 2024 Phys. *Lett. B* **861** *139267*

Remembering Hans Specht

Wolfram Weise

When I was a young postdoc and first met Hans Specht, he was already a renowned nuclear physicist, famous for his measurements confirming the existence of the "double-humped" fission barrier. In later years (during the 1990s), I was privileged to enjoy many exciting discussions with him. I always felt that these discussions were inspiring examples of productive communication between experiment and theory. Here, the brilliant experimentalist demonstrated the ingenuity of his CERES/NA45 experiment at CERN to the younger theorist who listened in amazement. On the theory side, we were following the basic ideas of spontaneously broken chiral symmetry and its interplay with the dynamical mass concept in QCD. In this context, one of the key questions was whether a dropping mass of the ρ meson in hot and dense matter would indicate a tendency towards chiral symmetry restoration. Theory groups in Japan, the US, and Germany (including ourselves) were actively pursuing this research in those days. Hans continually motivated and pushed us theorists to come up with predictions of the low-mass dilepton spectrum for which he and the CERES collaboration had provided pioneering precision data. Of course, the case turned out not to be as simple as the theoreticians had originally thought. The ρ meson has a large decay width, which increases in surrounding strongly interacting matter, so the primary in-medium effect was a flattening of the spectrum rather than just a mass shift. The deeper understanding and clarification of these phenomena were very much driven by Hans Specht's insistence and the quality of the CERES data (Fig. 4.20).

During the 1990s, I had various other opportunities to exchange with Hans at different levels. During his directorship at GSI, I could witness with great admiration how he conducted and promoted the cancer therapy project using ^{12}C beams, which culminated in the highly successful Heidelberg Ion-Beam Therapy Center. Later on, we met again in the founding Board of the European Centre for Theoretical Studies (ECT*) in Trento. His continued support for this European forum of exchange between theory and experiment was of great importance, especially in sometimes difficult periods. We had good times together whenever he visited the beautiful Villa Tambosi at the centre during the years I served as ECT* director.

Hans Specht was not only a brilliant physicist but also an excellent musician. Whenever we had a chance to meet, the discussions usually started with physics but sooner or later turned over to music. In later years, we discovered that both of us followed a ritual of beginning days at home on the piano with one of the preludes from Bach's Wohltemperiertes Clavier (of course, as a superior pianist, Hans always continued with one of the highly demanding fugues). Being a passionate scientist at the same time, he was not satisfied to just interpret his favorite Chopin pieces but

Fig. 4.20 Wolfram Weise's plenary talk at the Quark Matter 1996 (QM96) Conference in Heidelberg, organized by Hans Specht. © Hans J. Specht. All rights reserved

went on to do research on the neurophysiological basis of music perception in the brain. A key experience for me personally in early 2000 was when I listened to a colloquium he delivered together with Hans-Günther Dosch in Munich. They gave an exciting presentation of their joint interdisciplinary explorations, and these were then documented a few years later by unique articles in Nature Neuroscience.

A brilliant experimentalist and passionate researcher at the interface of physics and music, Hans Specht has left profound memories in my own scientific and cultural life. He will be missed.

Some Personal Memories of Hans Specht

Charles Gale

When Sanja Damjanovic told me about her plans to assemble a volume in the memory of Professor Hans Specht and asked me if I would contribute some words about my personal interactions with Hans, I gladly agreed. As I started to organize my thoughts, I realized I had first met "Professor Specht" before getting to know "Hans."

I was a young and very green Ph.D. student when I attended my first major physics conference: "The International Conference on Nucleus–Nucleus Collisions," held in 1982 at Michigan State University. That meeting had a special buzz to it, as it coincided with the inauguration of MSU's superconducting cyclotron facility. The atmosphere at that meeting was electric and crackling with optimism: a major new machine was coming online, several projects were in the planning stages, and everyone present was excited about the future. I can't claim I remember Hans' talk at that event but reading the typewritten proceedings (probably typed on the then ubiquitous IBM Selectric), I recognize his sense of history and of scientific opportunity. He spoke on "the possible experiments at ultrarelativistic energies" and on how measurements of the dilepton spectrum could reveal temperature and "provide a unique thermometer for the primordial phase."

My first genuine encounter with Hans was a few years later. In 1985, this nearly graduated doctoral student made his first trip to Europe to attend a summer school in Erice, Sicily, on "Nucleus–Nucleus Collisions from the Coulomb Barrier up to the Quark–Gluon Plasma." In addition to the excitement of discovering a new country and culture, there was the thrill of putting a face on each of the names I had read on the first page of many scientific papers. At that summer school, Hans gave a set of lectures on "Perspectives of Ultrarelativistic Nucleus–Nucleus Collisions." He spoke with excitement of the possible signatures of a phase transition in QCD, and of the planning of the NA34 ("Helios") experiment at CERN. His enthusiasm was infectious.

The Majorana Centre in Erice had a common room where attendees would gather at the end of the day's lectures to socialize and mingle. That room had a piano, and this is where I discovered Hans' second passion: music. I do not write "hobby," as I believe he was one of those individuals incapable of having a mere "hobby":

everything was pursued with the same intensity, at full throttle, like when he drove his Lotus, or when he spoke about Glenn Gould. Of course, the discussion topics during these social moments included physics but generally went far beyond. That meeting left a lasting impression on me. In addition to the exciting science I learned in the lectures, I met individuals for whom research was not a job but a way of life.

I ran into Hans more and more often as time went on and as our scientific interests grew closer. It seems that one of our regular meeting places was the ECT*, the European Centre for Theoretical Studies in Nuclear Physics and Related Areas, in Trento, Italy. One of those memorable encounters there was at a workshop on electromagnetic probes that Volker Koch, Krzysztof Redlich, and I co-organized in 1999. One of the central themes of the workshop was the theoretical interpretation of measurements by the CERES Collaboration (of which Hans was a key member) of low invariant-mass electron–positron pairs in heavy-ion collisions at the CERN SPS. Those results had generated much interest at Quark Matter '96, held a few years earlier in Heidelberg (and co-organized by Hans): they clearly signaled an enhancement of the dilepton signal over that expected from radiative decays or from appropriately scaled nucleon–nucleon reactions. That ECT* workshop provided an appropriate backdrop to lively exchanges, theatrical performances, and passionate debates. Hans was a sobering presence, reminding everyone to let the data speak and that too much room for speculation drives the need for more precise measurements. This was a prophecy soon to be fulfilled (Fig. 4.21).

Fig. 4.21 Front row (left to right): Charles Gale, Jörn Knoll, Hans Specht, Volker Koch, Dinesh Kumar Srivastava, and Ines Campo—ECT* Trento, 1999. © ECT* Trento. All rights reserved

At another ECT* workshop that I co-organized some years later, the results of the NA60 experiment (of which Hans was again a core member) were presented and immediately set a new standard. Those data represent to this day the most precise characterization of the spectral density of the low-mass vector mesons measured in high-energy heavy-ion collisions.

This string of successful experiments suggests another quality of Hans Specht, the researcher: persistence. Indeed, his drive and his devotion to many of the measurements of electromagnetic radiation in hadronic collisions over several decades are nothing short of remarkable. Our field owes a lot to his determination.

As I write this note, it dawns on me that I will have known Hans Specht for more than 40 years. The souvenirs I will keep are those of an extremely clever scientist, of a relentless researcher, of a charismatic—at times flamboyant—leader, and of a passionate human being. Our world is poorer without him.

Hans J. Specht: From a Purely Academic Association to Deep Understanding

Dinesh Kumar Srivastava

I was associated with the Variable Energy Cyclotron Centre, Kolkata from 1971 to 2019, having joined it through the Training School of the Bhabha Atomic Research Centre, Bombay (now Mumbai) in 1970, just after my graduation. In the first year we were given what we now call orientation courses on all aspects of nuclear sciences and assigned to various units and divisions. I was assigned to the Variable Energy Cyclotron Project, which was constructing a K130 room temperature cyclotron. We were expected to develop expertise in a given field and to have a good general idea of all fields in the Department of Atomic Energy. This was ensured by regular rigorous promotion interviews which tested our depth and breadth of understanding.

Thus, even though initially I was working in direct nuclear reactions like elastic and inelastic scattering of light nuclei and direct break-up of light ions, I was expected to know the latest developments in nuclear fission, because of the emphasis on nuclear energy. So it was that I read with great delight two excellent reviews about nuclear fission by Hans Specht [1]. I had also read with fascination about his observation of distinct rotational bands in plutonium-240. These had shown for the first time that nuclei can be in a strongly deformed cigar-shaped state shortly before fission, confirming the concept of a "double-humped" fission barrier.

The nuclear physics community during those early days was fascinated with the search for superheavy nuclei. Researchers were looking for them in rock samples, rocks from the Moon, monazite sand, etc. GSI offered the possibility of studying the electronic structure of superheavy elements by bringing two heavy nuclei rather close, with separations of the order of tens of fermis. This was minuscule compared to the orbital radii of electrons, offering an opportunity for the nuclei to rearrange themselves into configurations nearly corresponding to an element having a nucleus with neutron and proton numbers equal to the sum of those for the two nuclei. And this, well before such a superheavy element could actually be produced. I was fascinated by this possibility and remember reading everything I could find on the subject, even though it had nothing to do with what I was working on. Naturally, Professor Specht's name kept popping up in the references.

A little after 1980, a senior colleague of ours, Dr. S. S. Kapoor, visited the University of Heidelberg. Along with Professor Specht's group, he performed a very interesting and ingenious experiment which offered a "Direct Observation of Proximity Effects in Ternary Heavy-Ion Reactions." It studied two- and three-body exit channels in 12.4-MeV/u nucleus–nucleus collisions. The authors reported a high occurrence of three-body events, which arose primarily from two-step reactions, where large energy losses in the first step led to fission in the second. They also observed strong Coulomb proximity effects in the three-body final state which, treated quantitatively in Coulomb trajectory calculations, established a time scale of 1×10^{-21} s between the consecutive scission acts (scission to scission) [2]. This experiment was again discussed repeatedly in seminars.

Thus, Hans Specht was already a well-known name for me when our group started working on the quark–gluon plasma which was expected to be produced in relativistic collisions of nuclei. We started looking at the production of photons and dileptons as possible probes of the state of de-confined strongly interacting matter and its temperature. We were also fascinated and intrigued by the many predictions from various groups that the mass of ρ-mesons could be reduced due to the likely chiral symmetry restoration predicted to occur if such a state quark–gluon plasma was formed, and the consequences of such a reduction.

Our group performed the first calculation of thermal photons from an expanding, cooling, and hadronizing quark–gluon plasma undergoing transverse hydrodynamic expansion [3]. Most studies till then had neglected the radiation of photons from hadronic matter. Not knowing this rate of production, but knowing that there must be interactions in the hadronic matter before it undergoes a freeze-out, we made a seemingly reckless assumption that those rates are the same as that for quark matter at a given temperature. We were inspired by a similar assumption for dilepton production, which Hwa and Kajantie had made in their famous paper [4]. We concluded that the thermal production of photons could be modest at SPS energies, but large at low transverse momenta at RHIC and LHC energies. These were the pre-preprint days. However, we were soon to learn that Lichard, Kapusta, and Seibert [5] had actually calculated the production of photons from quark–gluon plasma and hadronic matter at 200 MeV and found them to be nearly identical. We heaved a collective sigh of relief.

We then calculated the production of thermal photons in S + Au collisions at CERN SPS energies [6] under two contrasting scenarios: one starting with a quark–gluon plasma, going through a mixed phase, then a hadronic phase, due to expansion and cooling, and undergoing a freeze-out; and the second assuming that there was no formation of a quark–gluon plasma and that the initial state was hot hadronic matter. The two scenarios had the same particle rapidity density, which was taken from experiments. The results for the scenario with quark matter in the initial phase were found to be similar to the preliminary data for single photons circulating at that time. Several of our assumptions were questioned and criticized. The data itself when finalized became upper limits, but one thing was becoming clear: a hot hadronic initial phase was unlikely at those collision energies.

One day around this time, I got a call from the director's office. I rushed over and found that Itzhak Tserruya from the Weizmann Institute had called. I knew his name from heavy-ion collision studies at low energies but had never met him. He congratulated me for the above work and asked me if I could perform the same calculation for dileptons. The CERES experiment was finalizing its results and they were seeing dilepton production above the background. The following month was a period of intense activity. I did not know how to put detector cuts into the theoretical calculations. Professor Tserruya patiently explained that to me. These interactions resulted in a deep friendship with him which continues to this day and has expanded to include our families and students who have gone to him for postdoctoral positions.

Up till then I had used only pion annihilation via the ρ meson to get dileptons from the hadronic phase. Charles Gale provided a more complete rate of production of dileptons from hadronic matter. Our results were in satisfactory agreement with the CERES data [7]. In our excitement, we completely overlooked the other interesting fact that, while the ρ peak appeared at its position for decay in vacuum, it was considerably broadened! Soon, however, we realized that this ruled out a large number of suggestions that the ρ mass decreases if there is quark–gluon formation and strongly supported the view that its width increases due to interactions in the medium.

While our paper was under consideration, Professor Hans Gutbrod organized for me to visit GSI for a period of about three months. That was the first time I met Hans Specht. He was the director at the time, and I made a courtesy call on the second day. Soon, we were in an intense discussion about our photon paper and the dilepton preprint. Indians traditionally tend to be deferential to seniors. I was no different. However, I soon realized that Professor Specht had thought of all possible scenarios, all possible sources of dileptons and photons, and all aspects of the dynamics. These discussions helped me revise our paper, which was published soon after.

The 1996 Quark Matter conference was to take place in Heidelberg. We had submitted our dilepton paper, which was accepted for oral presentation. However, I had no funds for travel, local expenses, and the registration fee. I wrote to Itzhak Tserruya and Jean Cleymans about my dilemma and situation. They encouraged me to ask the conference for support. I came to know years later that Professor Specht had personally intervened to provide funds for my participation. I remember that conference for two more reasons. First, it only had plenary talks and poster presentations providing sufficient opportunities for in-depth discussions with students. Second, as

I was late in booking a hotel, I was unable to get a room in any of the hotels with the available funds, and Jean Cleymans kindly offered to share a room with me in a modest hotel I could afford, foregoing his own upscale hotel (Fig. 4.22).

The medium modification of ρ-mesons originally seen in the CERES experiment and further detection of dileptons from the quark–gluon plasma in the NA60 experiment proved to be crucial in establishing the existence and properties of quark–gluon plasma, and the enduring quality and relevance of these measurements remain unsurpassed almost two decades later.

I met Hans again briefly at ECT* Trento in 1999, where a long program had been organized by Charles Gale to discuss electromagnetic probes, and my admiration for Hans' single-minded pursuit of thermal dileptons was reinforced.

Hans Specht and Sanja Damjanovic visited India during the Quark Matter 2008 conference at Jaipur. Being a perfectionist, Hans' continued to brief Sanja about presenting her results, right up to the last moment. The talk by Sanja and her conclusions essentially summed up the results of the passionate pursuit Hans had devoted himself to, single-mindedly and with clarity (as he used to say, "to the best of my insight"):

- Pion annihilation makes a major contribution to the lepton pair excess in heavy-ion collisions at SPS energies in the region $M < 1$ GeV.
- The in-medium ρ spectral function is identified, with no significant mass shift of the intermediate ρ, only broadening. Is there a connection with chiral restoration?
- First observation of the radial flow of thermal dileptons, providing a mass dependence tool to identify the nature of the emitting source. Is the radiation mostly partonic for $M > 1$ GeV?

Fig. 4.22 The author presenting his work [7] at QM'96 (Photo courtesy Sanja Damjanovic). © Hans J. Specht. All rights reserved

Fig. 4.23 Hans Specht (right) in discussion with Santoor Maestro Pundit Shiv Kumar Sharma (left) during QM2008, Jaipur. © D.K. Srivastava. All rights reserved

I remember this visit because I discovered yet another dimension of Hans' personality when he discussed the finer points of Indian classical music with one of our most respected Santoor players, Pundit Shiv Kumar Sharma (Fig. 4.23).

I met Hans Specht briefly when I was on a very short visit to CERN in 2016, and I found that, despite his advancing age, he had not lost any of his passion and zeal in pursuing dileptons from the quark–gluon plasma!

No wonder that Hans Specht will forever be remembered as a brilliant scientist and experimental physicist with a keen eye for cutting-edge detector concepts and a clarity of thought and single-minded devotion to details and accuracy. Looking back, I recall that, while he used to question all my assumptions during discussions, he never once questioned my assumption of the formation of quark–gluon plasma in the relativistic collision of heavy nuclei at SPS energies!

I am truly grateful to Sanja Damjanovic for giving me this opportunity to pay my respects to him.

References

1. "Nuclear Fission", Hans J. Specht, Rev. Mod. Phys. 46 (1974) 773; "Spectroscopic properties of fission isomers", V. Metag, D. Habs, and H. J. Specht, Physics Reports, 65 (1980) 1
2. D. v. Harrach, P. Glässel, L. Grodzins, S. S. Kapoor, and H. J. Specht, Phys. Rev. Lett. 48 (1982) 1093; P. Glässel, D. v. Harrach, H. J. Specht, Z Physik A 310 (1983) 189–216
3. J. Alam, D. K. Srivastava, B. Sinha, D. N. Basu, Physical Review D 48 (1993) 1117

4. R. C. Hwa and K. Kajantie, Phys. Rev. D 32 (1985) 1109
5. J. Kapusta, P. Lichard, and D. Seibert, Physics Review D 47 (1993) 4171
6. D. K. Srivastava and B. Sinha, Phys. Rev. Lett. 73 (1994) 242
7. D. K. Srivastava, B. Sinha, C. Gale, Phys. Rev. C 53 (1996) 567

Hans Specht—Thermal Radiation from the Quark–Gluon Plasma, and Much More

Gianluca Usai

In this brief text, I would like to honor the memory of Hans Specht and his remarkable contributions to the field of heavy-ion physics, particularly in the measurement and understanding of thermal radiation produced in high-energy nuclear collisions. This radiation, consisting of real and virtual photons, provides crucial insights into the temperature of the medium and signals of chiral symmetry restoration in strong interactions. Virtual photons, in particular, can be detected through their decays into lepton pairs. The story begins with an overview of the measurement of muon pairs, since it was the NA60 experiment which, in the early 2000s, first measured thermal muon pairs with unprecedented precision. Hans's contributions to this experiment were truly fundamental.

The NA60 experiment was a follow-up to the previous NA50 experiment at the CERN SPS. The apparatus was designed to measure muon pairs produced in nuclear collisions, utilizing a muon spectrometer inherited from NA50. This spectrometer consisted of a thick hadron absorber to stop hadrons, a toroidal magnet, and a system of tracking chambers based on MWPC. However, this apparatus had significant limitations. Multiple scattering in the hadron absorber degraded the mass resolution to approximately 70 MeV at the ω mass. Additionally, the hadron absorber severely limited the acceptance in transverse momentum, particularly for muon pairs with masses below 1 GeV.

In the late 1990s, radiation-hard silicon pixel detectors were developed for LHC experiments. Around this time, Peter Sonderegger, a brilliant physicist, proposed complementing the muon spectrometer with a "vertex spectrometer." This telescope, consisting of planes of pixel sensors immersed in a magnetic field, would measure all particles, including muons, before they entered the hadron absorber. C. Lourenço began collaborating with Peter and secured a staff position at CERN with the specific task of running NA60. I arrived at CERN in 2000 as a scientific associate and joined the project, tasked with designing the readout for the silicon tracker (and later for other components as well). I found a very small team and initially I was skeptical that we could ever run the experiment within the short timeline: a lead run was scheduled in 2002 and an indium run in 2003 (SPS was going to stop in 2005). The silicon tracker had to be built from scratch and a new, much faster readout was needed also for the muon system. A small team of brilliant and motivated Ph.D. and postdoc students was formed. In less than 2 years, under the coordination of Carlos and myself, this team handled the design and testing of silicon planes, the readout system, detector control system, and data acquisition.

The main motivation for the experiment was to measure the muon pairs in the so-called intermediate-mass region (IMR) between the ϕ and J/ψ mesons. NA50 observed an excess of muon pairs compared to the expected yield from Drell–Yan processes and simultaneous semi-leptonic decays of $D\overline{D}$ mesons. However, it was unclear whether this excess was due to an enhancement of the charm cross section or a prompt source like thermal radiation (from partonic and/or hadronic matter). The NA60 vertex detector would enable precise measurement of displaced muons originating from $D\overline{D}$ mesons, making it the key to solving the puzzle. On the other hand, it was initially unclear whether the ρ meson, considered a probe of chiral symmetry restoration in the hot medium, could be measured at lower masses due to the transverse momentum cut.

In 2002, Hans Specht joined NA60. I remember his first participation at the Lisbon collaboration meeting in early summer. While I was impressed by his vast knowledge of physics, I was even more amazed when I saw him swimming in the cold ocean, holding a thermometer that registered a chilling temperature of 12°. It was soon realized that, thanks to the magnetic field in the vertex region, many soft muons were recovered that would otherwise be outside the acceptance. For the first time, it became clear that the ρ meson could be measured. Hans played a fundamental role in reshaping the perspective of what NA60 could achieve.

The lead run planned for 2002 was canceled for technical reasons. After a commissioning test beam for the silicon tracker in August 2003, the indium run took place in October 2003. The run was a success, collecting a world-record statistic of approximately 5×10^5 reconstructed muon pairs, a record that remains unsurpassed more than 20 years on. The mass spectra immediately appeared astonishing. In the low-mass region, the ω and ϕ mesons stood out like giant towers, with a mass resolution of around 20 MeV, never before seen so clearly in heavy-ion collisions. Initial results were presented at numerous conferences, including Quark Matter and Hard Probes, in 2003 and 2004. However, there was still a long road ahead before quantitative science could be conducted with those spectra. The most challenging aspect of the data analysis was the subtraction of the combinatorial background from muons originating in pion and kaon decays. Additionally, muon tracks could be incorrectly matched to hadronic tracks in the vertex tracker, creating another source of background. It took nearly 2 more years to master the subtraction procedure and evaluate all systematic uncertainties. Ruben Shahoyan played a crucial role in this effort. Known for working tirelessly for days without sleep, he single-handedly wrote all the reconstruction software. The first results were presented during the collaboration meeting in Alghero, Italy, in spring 2005. Later that year, they were publicly presented at ECT* in Trento, Quark Matter in Budapest, and other conferences. I remember Jurgen Schukraft complimenting the NA60 team for the incredible "fireworks" when they presented their results at Quark Matter. By the end of 2005, after I began serving as spokesperson, the collaboration was ready to publish the scientific results.

The detailed characterization of thermal radiation was a major contribution by Hans and Sanja Damjanovic, who completed her Ph.D. in CERES. The excellent mass resolution allowed us to easily subtract the hadron cocktail of ω and ϕ resonances at freeze-out, isolating the thermal radiation produced by the ρ meson through $\pi\pi$ annihilation. The associated ρ spectral function, averaged over the space–time evolution of the fireball produced in a nuclear collision, shows a nearly diverging width as chiral symmetry restoration is approached, but essentially no shift in mass (Phys. Rev. Lett. 96, 162,302 (2006)). This resolved a decades-long controversy regarding the spectral properties of hadrons near the QCD phase boundary.

Thanks to the vertex detector, it was shown that the data in the IMR are perfectly consistent with no enhancement of open charm. The observed excess was isolated by subtracting the Drell–Yan and open charm contributions from the total data, and it was found to be fully consistent with thermal radiation (Eur. Phys. J. C 59 (2009) 607). The exponential shape of this excess, resulting from a nearly flat spectral function, obeys $dN/dM = M^{3/2}\exp(-M/T)$, where T is a space–time average over the system's evolution. A fit of the mass spectrum with this expression over the range $1.2 < M < 2$ GeV yielded $T = 205 \pm 12$ MeV (AIP Conf. Proc. 1322 (2010) 1). Since mass is a Lorentz-invariant quantity, the mass spectrum is unaffected by the motion of the emitting sources, unlike transverse mass spectra. The parameter T is therefore purely thermal, and the fit value of approximately 200 MeV, significantly above the pseudo-critical temperature $T_c = 155$ MeV, implies that partonic emission dominates in the IMR. This conclusion was corroborated by analyzing the effective temperature of the transverse mass spectra as a function of mass. Further insight was gained from the analysis of angular distributions, which showed that thermal radiation is not polarized (Phys. Rev. Lett. 100, 022302 (2008); Phys. Rev. Lett. 102, 222301 (2009)).

Without a doubt, Hans had a profound influence on me, both as a researcher and as a person. I remember many occasions where he expressed his "credo"—a term he deliberately used in Italian—of uncompromising excellence. He was undoubtedly a strong character and a free spirit, with an incredible attention to detail, no matter how small. Our discussions on physics were endless, and he would constantly argue until he was fully satisfied on every aspect.

Over the years, we attended many conferences and workshops together. Among these, I particularly remember the meetings organized at ECT* in Trento. He loved the atmosphere of the small town, bustling with students, especially in spring and summer. We often walked downtown and enjoyed chocolate ice cream with orange— a flavor we both adored. Even as he neared the end of his life, he called me to ask when we would organize another meeting in Trento. Unfortunately, we will no longer have the opportunity to share those delightful little workshops together.

Beyond science, where he was a true master, Hans had a genuine interest in many other aspects of life, many of which we shared. While at CERN, I often visited him, and we would spend hours discussing physics and many other topics. He was, in his own words, "crazy" about the sea, and we bonded over this shared passion, especially since I live on an island in the middle of the Mediterranean. He loved watches, and we often discussed this shared interest, as we both owned an Omega Speedmaster—the iconic moon watch. Then there was music. My father, who shares Hans's age and a fanatical passion for music, had a similar character. This likely made Hans feel even more familiar to me. Hans was also well known for his passion for fast cars. He could often be seen driving his Lotus around CERN. Once, he told me that he had received a letter from the Swiss government banning him from driving on Swiss motorways due to the excessive number of speeding fines. On another occasion, during a workshop in Germany, I couldn't take a train to the airport due to a strike. Hans immediately offered to drive me in his BMW, and I remember him speeding at over 250 km/h on the motorway to ensure I made my flight.

The legacy of NA60 lives on in the NA60+ proposal, which aims to perform precise measurements of thermal radiation and charm at high baryochemical potential through a beam energy scan at the CERN SPS. When NA60 came to an end, the collaboration between myself, Ruben, and Hans continued. Together, we shaped the concept of NA60+ between 2010 and 2014. Later, Enrico Scomparin and others joined the effort, and we began collaborating to bring the experiment to life. As of today, the project is under discussion by the CERN SPSC, with a timeline to start data collection in 2030.

Before concluding, I would also like to remember Louis Kluberg, another key figure in the NA60 experiment. He was the former spokesperson of NA50 and renowned for his work on J/ψ suppression. Hans and Louis, similar in age, shared strong ideas and a deep passion for physics. During collaboration meetings, they constantly exchanged this fervent enthusiasm.

Hans, though you are no longer with us, your legacy endures. I will miss you deeply, but your influence will remain as a part of me for the rest of my life.

Brilliance in Dilepton Radiation

Ralf Rapp

I had the privilege of rather numerous interactions with Professor Hans Specht, always enjoyed them thoroughly, and benefitted deeply from them. All of these were, in one way or another, in the context of our common professional passion for dilepton radiation in high-energy heavy-ion collisions. Professor Specht was a true pioneer in this field and made several groundbreaking discoveries. I am sure that other colleagues and friends can do better justice to comprehensively characterize his enormous achievements, and therefore I will focus on my personal perspective.

The first direct encounters I had with Professor Specht were in the context of the dilepton data taken at the SPS with the CERES detector for a Pb beam at 40 AGeV bombarding energy. This was in the late 1990s, toward the end of his directorship at GSI. These data followed in the aftermath of the breakthrough discovery of thermal radiation of low-mass dileptons at full SPS energies using 158 AGeV lead and 200 AGeV sulfur beams. There was much discussion on what to expect at lower beam energies, and one of the attractive perspectives was that the signal would be smaller, as a consequence of a less hot and shorter-lived quark–gluon plasma.

However, Professor Specht was, in his own words, "only interested in the truth" and did not compromise on facts. Paired with his experimental ingenuity and relentless desire to search for precisely the truth, this was an unbeatable combination. At the end of the day, the relative low-mass dilepton enhancement in the 40 GeV Pb run at the SPS turned out to be even larger than at the higher energies, thereby providing key indications for the fundamental mechanism of chiral symmetry restoration in hot and dense QCD matter.

The CERES results already constitute a major milestone in heavy-ion physics, but they were followed by even greater discoveries, which many still consider the "gold" standard in thermal dilepton radiation to this day (even though it was achieved with indium beams). I am, of course, referring to the fantastic dimuon data taken in In-In collisions at the SPS by the NA60 collaboration. The original motivation was to look for J/ψ suppression in an intermediate-size collision system, and the pertinent NA60 finding very significantly propelled forward our understanding of this important physics. However, the discoveries associated with the thermal radiation signal arguably outshone the former.

I clearly remember discussions with Hans Specht and his co-workers (especially Sanja Damjanovic) at one of the early Hard Probes meetings, held in Ericeira (Portugal) in 2004. Once again, I got a taste of the enormous challenges involved in extracting the dilepton signal from heavy-ion collisions, and at times I wondered whether my calculations, into which I had already invested a significant part of my life, would still be useful in the future. The experimental results that finally emerged due to the relentless effort and brilliance of Professor Specht and his co-workers were simply stunning: in the "low-mass" region (below ~ 1 GeV), an unprecedented measurement of the in-medium rho-meson spectral function emerged which subsequently enabled detailed insights into the mechanism of chiral symmetry restoration, and also directly reflected the very large interaction rates in the hot QCD medium, not unlike the strongly coupled quark–gluon plasma that was subsequently produced and studied at RHIC and the LHC. In addition, a measurement of the dilepton continuum at "intermediate" masses (above 1 GeV) allowed for a quantitative characterization of the early temperatures in the fireball which unequivocally demonstrated that the radiation originates from a quark–gluon plasma.

With additional measurements of transverse momentum spectra, the blue shift of the dilepton sources due to the collective expansion of the fireball could be assessed; these data fully corroborated that the intermediate-mass radiation emanated from early phases without significant blue shift, while the low-mass radiation mostly came from later times, consistent with large contributions from temperatures around the pseudo-critical transition temperature of chiral symmetry restoration. When reflecting on these breakthroughs some 20 years later, I am still amazed by the quality and scope of these magnificent achievements!

I also remember quite well the many discussions that were held at various workshops at the time, in particular at the famous ECT* venue in Trento. While these discussions could sometimes get heated, I always found them highly constructive and most insightful. Among the traits of Professor Specht that I most appreciated was his keen interest in communicating with his theoretical colleagues, driven by the desire to maximize the physics insights that could be gained from his data. At the same time, he remained completely impartial with regard to the intense competition between the various theory groups working on the various aspects of dilepton production.

I certainly owe a good part of my knowledge and understanding of QCD matter formation in heavy-ion collisions that emerged from the NA60 dileptons to discussions with Professor Specht and his team. I can easily see how he was an extraordinary mentor to his students and postdocs, and how his training of the next generation of scientists in the field led to new leaders and valued colleagues, including Axel Drees and Thomas Ullrich here in the US. Indeed, many of Professor Specht's outstanding qualities have become benchmarks in shaping my own approach and efforts to research and training of the next generation. His positive attitude and the enthusiasm that he projected on me in our many discussions had an impact on me that went beyond the scientific realm.

While most of the time our conversations were centered on the most recent dilepton results (both experimental and theoretical), we did have the opportunity, although unfortunately somewhat limited, to delve into another passion that we shared: sports cars, in particular Italian ones. As usual, Professor Specht had clear and compelling opinions on this and what it meant to own and operate such a vehicle, and this certainly contributed to the fact that I still indulge myself by driving one today.

Professor Specht's legacy lives on, and his brilliant dilepton measurements will remain a highlight in the history and textbooks of QCD matter discoveries for decades to come, maybe longer. But I think it is also important to remember his straightforward personality that has been at the heart of his integrity and passion for science. Let me finish by recalling one of his favorite words: "Incredible!" And incredible he and his achievements have been. I will miss him.

Hans Specht—A Charismatic Leader

Volker Koch

Elementary particles like the proton or the pion are composites made from quarks and gluons, the fundamental building blocks of matter. The theory of the strong interaction predicts that, at very high temperatures and densities, matter is de-confined, meaning that the elementary particles disappear, and matter consists of quarks and gluons. But what happens at lower densities and temperatures? Do the particles change their mass, as a simple interpretation of the chiral symmetry of QCD might suggest? Or do they slowly melt away, first mixing with other states before they lose their identity? This fascinating question about the in-medium properties of elementary particles has been, and still is, at the heart of strong interaction research.

4 Reflections and Tributes

This research received a major stimulus in 1995 with the first results reported by the CERES collaboration under the leadership of Hans Specht. CERES reported an unexpected enhancement in the measured dilepton mass spectrum, and this observation, although it contained large uncertainties, triggered more than 10 years of intense activity in the community. Was this a sign that the mass of the ρ-meson was reduced in the hot and dense matter? Or did this enhancement come from mixing with other states?

It was once again Hans Specht who provided the definitive answer. He realized that another experiment, NA60, originally designed to look for the suppression of the J/ψ particle, could be used to carry out a dilepton measurement with unprecedented precision. And in 2005, NA60 reported the results which showed unequivocally that the mass of the ρ-meson is not changed, but rather broadened, as expected if the ρ-meson mixes with other states.

During those years, I had the privilege of many, often very intense and at times even emotional discussions with Hans Specht during a series of the most stimulating workshops at the ECT* in Trento. I got to know a charismatic physicist with a keen intellect, who defended his convictions with passion and vigor. And while his strong personality often appeared to dominate the discussions, he was always open to arguments and immediately changed his opinion if the evidence demanded it.

One example was an evening discussion at the Trento Meeting in 2010 on how to extract the temperature of the fireball created in these collisions from a dilepton measurement. During that meeting Hans Specht showed the results of transverse momentum spectra and argued forcefully that their slope should measure that temperature, while others disagreed because of the blueshift due to collective expansion. So, we decided to have a dedicated evening discussion on the controversy. Not surprisingly things got pretty intense (maybe helped by the excellent drinks the ECT* provided), but did not result in any tangible result or agreement. However, the discussion stuck with Hans Specht and about a week after the workshop we received an email from him in which he agreed (a) that the transverse momentum spectra were no good and (b) that one should instead use the invariant-mass spectra at intermediate masses, a suggestion which had come up during our evening discussion. As a result, he and the NA60 collaboration were the first to extract a lower limit for the temperature of the fireball which turned out to be above the so-called deconfinement temperature, thus indicating that a quark–gluon plasma was indeed created in these collisions.

In later years, I learned during less emotional conversations that Hans Specht had many other interests such as music and that he even published papers on music and the brain. I also learned that he cared deeply about the health and future of the physics community, in Germany in particular.

I will remember him as a unique character and charismatic leader who will be sorely missed.

Physics and Music

Hans Günter Dosch

H. J. Specht was not only a lover of music and an excellent pianist but also, and how could it be otherwise with such an enthusiastic physicist, productively interested in the connection between physics and music. In his biography in the series "Emeriti remember [1]," he devotes an entire section to "Music, Physics and Neurophysiology" and writes that this is "THE hobby in his professional life," adding: "with a strange beginning around 1985." The strange beginning came about during a follow-up session to the Friday colloquium, where we were both discussing a possible lecture series on this topic. The forthcoming 600th anniversary of the university provided an excellent occasion for this, because one of, if not THE, key works on the physiological basis of the laws of music, "Die Lehre von den Tonempfindungen" [2] by Herrmann Helmholtz, had been published during his time in Heidelberg. We therefore announced the 2-h per week lecture "Helmholtz und danach, Physik und Musik" as a contribution to the anniversary year for listeners from all faculties.

The integration into the anniversary year quickly had a positive effect. Our colleague Joachim Heintze, also an enthusiastic music lover, had saved a grand piano that the Bechstein company had donated to the university on the occasion of its 500th anniversary from being scrapped. However, it was in a desolate state and although the financing of the renovation was quite moderate for an experimental institute, it had to be justified to the administration in view of possible criticism from the Court of Auditors. The public anniversary lecture was of course ideally suited for this justification, especially since HJ Specht, as an enthusiastic and outstanding pianist, was able to give spontaneous audio examples during the lecture. The grand piano is still in the lecture hall today and is used time and again for public events on the subject of "Physics and Music."

In the lecture, there was a clear division of tasks between theory and experiment, but not a strict separation, because although we both completely agreed on our love of music and physics, we sometimes disagreed on details. Some people thought that we had carefully rehearsed our mutual teasing, but we were not actors, we were just expressing our opinions bluntly. We, and I think the listeners too, had a lot of fun in the process.

In the lecture, we paid particular attention to the physiological foundation of musical consonance and dissonance. The first, decisive approaches to this field go back to Helmholtz [2]. In the more recent discussion, the so-called critical bandwidth [5] plays an important role. If two sounds are perceived simultaneously and their tonal distance is less than a critical bandwidth, they begin to merge and create a dissonant or harsh sound impression; this effect can be explained by the structure of the inner ear.

In the middle and higher registers, the critical band width corresponds to an interval between the minor third and the whole tone; the thirds themselves are important harmonic chords in polyphonic music. In the lower register, however, this band width

4 Reflections and Tributes

from F. Chopin; Scherzo II Op. 31

L. van Beethoven Sonate für Klavier (Apposionata) Op. 57 end

Fig. 4.24 Several bars from Scherzo II by Chopin and the Appasionata by Beethoven. While Chopin avoids thirds and even fifths in the basses, Beethoven uses them as special accents

increases and therefore a third in the lower register already has a thoroughly dissonant character. Therefore Chopin avoids thirds in the lower registers, whereas Beethoven deliberately uses them as accents, for instance at the end of the Appasionata; this can be clearly seen in the example 1. Naturally, this fascinated the pianist Specht, who loved Beethoven and Chopin, and he never missed a suitable opportunity to play these or similar examples on the piano (Fig. 4.24).

However, some statements in the literature about the increase in the critical band width seemed exaggerated to us, and this led to our first experiment in this field. We invited listeners to take a test in the large lecture hall one evening and, with 89 test subjects, we were able to achieve a statistical reliability that we rarely found in the relevant literature. In fact, our results confirmed a significant increase in the critical bandwidth at low tones, but it was much smaller than that published in some of the publications which arouse our suspicion.

The questions about critical bandwidth were quite topical, as they play an important role in the compression of digital acoustic recordings, for example when condensing to MP3 format. Nevertheless, we stuck to our guns as particle physicists and did not venture into further original research on the subject of music and physics (Fig. 4.25).

That changed a few years later. On occasion of the 100th anniversary of Helmholtz's death, we held in the winter term 93/94 another lecture series on the physical foundations of music. Among our listeners was a physics student who out of love for music had interrupted his physics studies to finish his diploma in church music. He had already been made aware of our lecture as a physics student in Freiburg by Martin Specht, the son of H. J. Specht, and was therefore one of our most eager listeners. He is not only an excellent church musician and organist but also has a

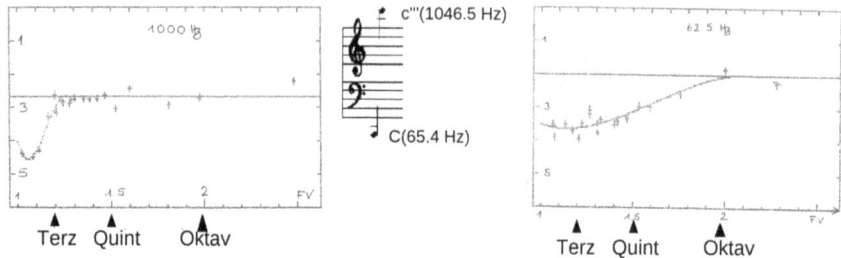

Fig. 4.25 Critical bandwidths at a high pitch, center frequency 1000 Hz, close to c''' and at a low pitch, center frequency 62.5 Hz, close to C, see example: two sine tones were played simultaneously with the frequencies f_1 and f_2 with the mean values $\sqrt{f_1 \cdot f_2} = 1000$ and 62.5, respectively and the frequency ratio $FV = \frac{f_1}{f_2}$. The task for the listeners was: rate the auditory impression of the "chord" with the scores 1–5; 5 for very rough, 1 for completely smooth. The mean values of the ratings of 89 participants are given here. © Hans J. Specht. All rights reserved

phenomenal ear: when other people talk about the whistling of a fan, he talks about the high F sharp in the corner.

One topic that interested us both very much, and which evoked also great interest in the lecture, was that of fundamental tone tracking. This topic goes back to the nineteenth century [3, 4], but was revived in particular by the work of Jan Frederic Schouten [6]. Schouten recognized that the interpretation of pitch, as advocated by Ohm and Helmholtz [2, 4], was not always valid. According to their theory, the fundamental tone, i.e., the tone with the frequency f_0 of a harmonic tone which is composed of the sine tones of the frequencies f_0, $2f_0$, $3f_0$ ... determines the pitch. The composition of the other sine tones creates the impression of timbre. Using sophisticated electronic methods, Schouten was able to show that a tone is perceived with the pitch corresponding to the frequency f_0 even if this "fundamental tone" does not occur at all in the harmonic tone. Today you can measure the spectrum of a tone with any better mobile phone and see this effect for yourself.

Even for a large concert grand piano, the fundamental tone is missing in the lowest notes. The spectrum of a Steinway B 211.1 (length 211 cm) can be seen in the following illustration (Fig. 4.26).

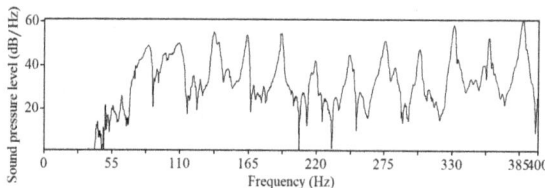

Fig. 4.26 The spectrum of the lowest note of the grand piano by H. J. Specht. It is the Steinway B211.1, one of the most played private instruments by professional pianists. The fundamental of the low A, 27.5 Hz, and the first overtone at 55 Hz are completely missing. © Hans J. Specht. All rights reserved

4 Reflections and Tributes

Peter Schneider was enthusiastic about this topic and developed as part of his diploma thesis—now also in physics—very detailed test procedures to refine the concept. If the fundamental tone alone is removed from a harmonic tone with many components, almost everyone will perceive an unchanged pitch and only notice changes in timbre. If several of the lower components are missing, however, the listeners are divided. Peter Schneider quantified these differences by special test examples and completed his diploma thesis on the subject of "Residuals and interval recognition of harmonic complex tones" under our supervision in 1996. As an illustration, a picture, Fig. 4.27 from a later paper [8] is presented here. A test for distinguishing between listeners who tend to track the (missing) fundamental tone from the entire spectrum is shown in Fig. 4.27, left, "Pitch test." Of two harmonic tones, the two lowest two components of Tone 1 and the lowest three components of Tone 2 are omitted. The fundamental tone of Tone 1 is higher than that of Tone 2, although the individual components of Tone 2 are higher or equal than those of Tone 1.

A pronounced fundamental tone listener (f_0) will therefore say that Tone 2 is lower, although this is not true for the individual components of the existing spectrum, whereas a listener who is more oriented towards the components (f_{SP}) will perceive Tone 1 as lower. In this way it was possible to form a very differentiated picture of hearing characteristics, from the extreme fundamental listener, at -1 in the graph, to the extreme spectral listener, at $+1$.

The very differentiated picture of different perceptions for the same sensory stimulus can be seen in Fig. 4.27, right, and is of course a great challenge for sensory

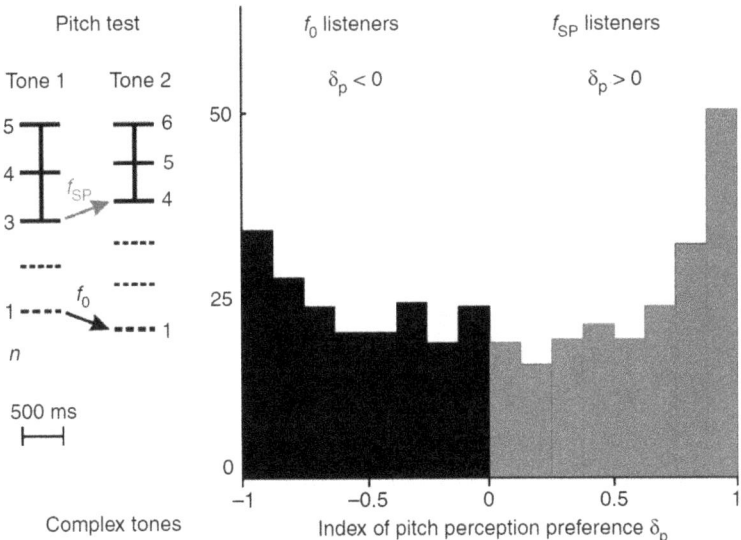

Fig. 4.27 Left: A test to differentiate between fundamental listeners (f_0) and spectral listeners (f_{SP}). Right: Distribution of the two groups depending on the relative strength; (fundamental: black; spectral: grey). From [8]. © 2005, Springer Nature America, Inc. All rights reserved

physiology: What are the causes of this differentiation, in particular whether it is determined more by environmental influences, e.g., training, or more by physiology. Clearly the psycho-acoustical tests which formed the basis of the investigation could not answer these questions.

Fortunately, at this time (1996), M. Scherg set up a laboratory at the neurological university hospital in Heidelberg and installed a magnetoencephalograph (MEG), a device that allows to observe physiological processes inside the human brain: It measures the currents in the human brain non-invasively, i.e., from the outside. We developed a harmonious and productive working relationship with M. Scherg, as well as later with his successor A. Rupp. Since we physiscists mainly used this device, I shall give some more details on it.

There are three reasons why such a measurement of processes in the brain is not easy:

(1) The currents in the brain have their origin in the flow of charges whereby the individual neurons communicate with each other. The resulting currents are small and only flow over short distances, the magnetic dipole momentum is typically 20 fA m for one neuron, with typically 1,000,000 neurons contributing to a signal that amounts to as much as 20 nA m, and the resulting magnetic field has a strength which is about 1/100,000,000 (10^{-8}) of the strength of the earth's mean magnetic field. The device must therefore be very sensitive and very well shielded against magnetic interference. It goes without saying that H. J. Specht, with his penchant for pushing the boundaries of what can be measured, was enthusiastic about such an instrument.

In the MEG, so-called SQUIDS (superconducting quantum interference devices) are used to measure the magnetic field. They are sensitive to the change of a magnetic flux quantum, which is proportional to the Planck quantum h and is correspondingly small, about 0.0001 fTcm^2. Therefore it is well suited to measure the magnetic fields produced by the currents in the brain. Figure 4.28 shows an MEG with a test subject in the magnetically shielded room. Since the SQUIDs are based on supra-conducting conductors they must be cooled with liquid helium.

(2) The next difficulty is that the signals are not only very small but also very noisy. Even the idle brain produces a lot of signals and the interesting ones have to be separated. This is only possible with a large number of measurements. We were both used to this from high-energy physics. Since the individual tones examined can usually be very short, repeating them many times is in most cases not a problem.

(3) The third difficulty is of theoretical nature: there are theorems of mathematical physics which state that it is impossible to draw unambiguous conclusions about the current distribution inside the head from the magnetic fields outside the head. The challenge is therefore to keep the uncertainties as small as possible by a clever selection and evaluation of the measurements.

The MEG makes it possible to gain insight into certain physiological processes in the brain; the individual anatomy of the brain can be determined very precisely with

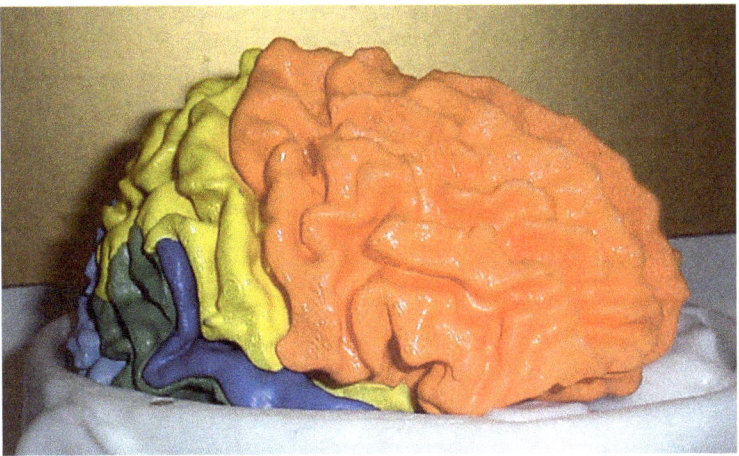

Fig. 4.28 3-D print of the coordinates of a human brain as recorded by MRT. The auditory cortex in the picture is shown in blue. © Hans Günter Dosch. All rights reserved

the help of magnetic resonance imaging (MRI). Figure 4.28 shows a 3D printout of a brain based on an MRI analysis. The primary auditory cortex, which is responsible for early sound sensation is colored blue; it is composed of a ridge-like elevation which is called Heschl's gyrus and the tilted axial section through the superior temporal plane of the human brain, the "planum temporale."

In Fig. 4.29 the results of the neurophysiological measurements (MEG) and the anatomical results (MRT) are shown. The test persons were 12 professional musicians, 12 amateur musicians, and 13 non-musicians. Already from the figure one sees the remarkable difference between the groups. In Fig. 4.30 the correlation of the physiological and anatomical parameters and with the musical aptitude is shown.

It is impossible to go into the details of the results here, but the strong influence of an anatomical trait such as the size of a particular part of Heschl's gyrus naturally raises the explosive question: disposition or environmental influence. In the case of musical aptitude seems to be decided that disposition plays at least an important role. Here the fierce criticism of some referees set in. However, the influence of the time spent practicing music on a daily basis could not be proven for the early components and it is also known from studies that the Heschl gyrus is stable in humans from the age of 7.

In his work, Peter Schneider also dealt in great detail with the question of fundamental tone recognition. There was a large number of test subjects: 125 practicing professional musicians, 181 graduate students of music, 66 amateurs, and 48 non-musicians. A subset of 51 professional musicians, 16 amateur musicians, and 20 non-musicians took part in MEG and MRI examinations.

Here also a distinct neurophysiological and anatomical signature for the fundamental and spectral listeners showed up, this time specified by a dominance in

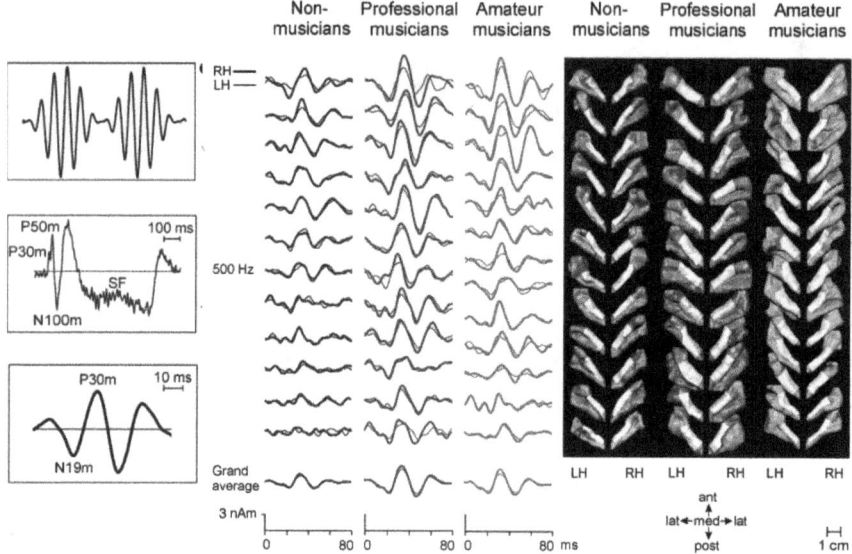

Fig. 4.29 Left: The auditory stimulus and evoked magnetic fields. The auditory stimulus was modulated with a frequency around 30 Hz, allowing to measure the reaction on the tone onset about 60 times per second. The averaged response shows the components P30m and P50m, N100m and the sustained field. After the deconvolution of the signal the typical early N19m- P30m signal appears. Right: The evoked N19m-P30m signals and the gray matter surface reconstruction for all test persons aligned in the same order. LH (thin lines) and RH (thick lines) indicate signals from the right and left hemisphere, respectively. From [7]. © 2002, Springer Nature America, Inc. All rights reserved

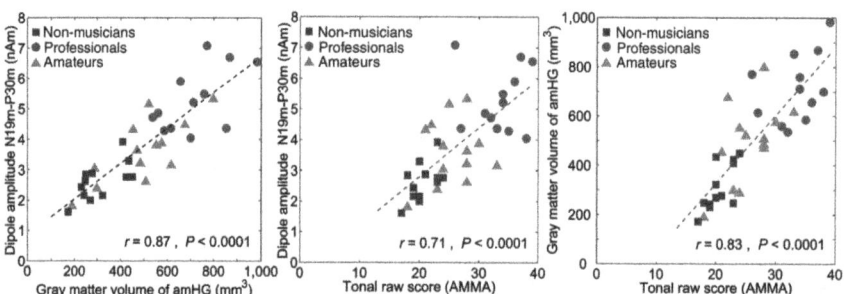

Fig. 4.30 Relation between neurophysiological, anatomical and the musical aptitude as measured by the well established AMMA test. The correlations are highly significant. From [7]. © 2002, Springer Nature America, Inc. All rights reserved

the right-hand half of the brain for the fundamental listeners (f_0) and a left-hand dominance for the spectral listeners (f_{SP}). For details see the caption of Fig. 4.31.

P. Scheider was especially interested in the influence of hearing preferences (degree of f_0 and f_{SP} respectively) on the choice of the musical instrument. H. J. Specht was a solid fundamental tone listener (f_0), as it is typical for classical pianists.

The collaboration between the Physics Department and the Neurophysiology Department of the Neurological Clinic was successfully continued and led to several further doctorates in physics. However, as they are less directly related to music than those discussed above, they are not discussed here and only the references are given at the end in the bibliography [9–13].

Especially after the lecture in 1994, which was also associated with the Helmholtz Conference of the Faculty of Medicine, many invitations to colloquia in Germany and abroad followed. Highlights were certainly the invitations to the Loeb Lectures at Harvard University in 1999 and to the supporting program of the music festival in Verbier. The colloquium at Cern is available as video on the Internet [14].

The external colloquia were held in the same elaborate style as the lecture in Heidelberg[1] where we had several evenings and nights to prepare for each lecture; there we had the competent support of the lecture assistant H.G. Siebig. This was of course not the case at the external events and the preparation on site was correspondingly hectic, especially when one has to take H. J. Specht's love of detail and perfectionism into account.

In Verbier, this meant that we were hardly able to make use of our right to free access to all events and rehearsals of the festival.

In Cambridge, MA, the secretary of the Loeb Foundation pointed out to us that the foundation was very rich and that we did not need to be restrained when eating in restaurants. But we normally got some fast food from the neighboring MacDonald because we could eat that during the preparation of the lectures. When we came to her with our final bill for the expenses, she said shaking her head: "I think we'd better do a per diem account."

But we were rewarded for our asceticism by the audience. We insisted that, as usual, we needed 90 min for the colloquium's tense topic: "Musical Harmony, Physics, Physiology, and Psychology." Our colleagues from the physics department made it clear that this was unusual and that many members of the audience would leave early. As this did not happen, a colleague told us: "You have glued them to their seats."

A worthy conclusion to our series of colloquia was our contributions to "Töne, Klänge, Musikalische Harmonie" at the International Academy Traunkirchen, organized by Anton Zeilinger in 2013 (Fig. 4.32).

It is a pleasure to thank Adelheid Specht, Peter Schneider, André Rupp, and Martin Messmer for their valuable information.

Most of all I thank H. J. Specht posthumously for many many hours of discussion, sometimes turbulent, but always stimulating.

---

[1] Only the mutual teasing was a bit milder than in the home games.

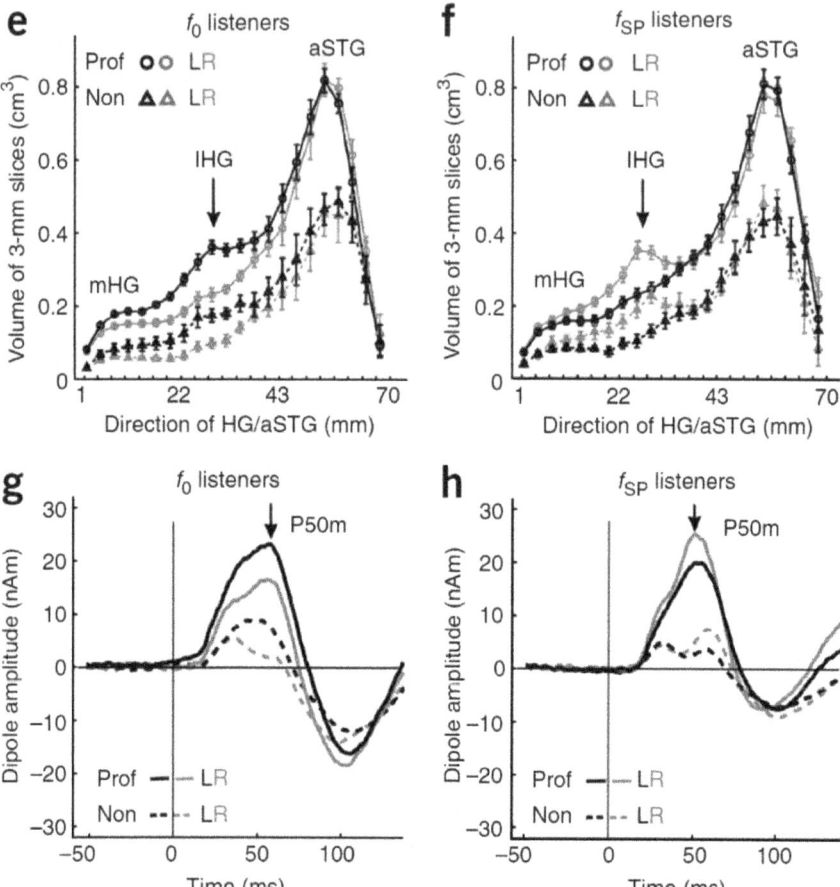

Fig. 4.31 General notation: right hemisphere: black; left hemisphere: gray. Professional musicians: Circles o (**e, f**) or solid lines (**g, h**); non-musicians: triangles Δ (**e, f**) or long dashed lines (**g, h**). In (**e, f**) the volume of a 3 mm slice of the Heschl gyrus is plotted along its length, one clearly sees the dominance of the right hemisphere for fundamental listeners (f_0) in the region at around 30 mm, whereas for spectral listeners the left hemisphere is dominant in the corresponding region. The difference between the hemispheres is larger for the professional musicians than for the non-musicians. In (**g, h**) the dipole amplitude around 50 ms after tone unset is displayed. Also here the right hemisphere is dominant for the fundamental and the left hemisphere for the spectral listeners, the large difference between the signals of professional musicians and non-musicians is remarkable. From [8]. © 2005, Springer Nature America, Inc. All rights reserved

4 Reflections and Tributes

Fig. 4.32 Participants and lecturers at the workshop *"Physik und Musik"* of the Akademie Traunkirchen, Austria (September 2013). Right: Anton Zeilinger, Nobel Prize with Alain Aspect and John F. Clauser for experiments with entangled photons. Middle: Peter Grünberg, Nobel Prize with Albert Fert for their discovery of giant magnetoresistance. © Internationale Akademie Traunkirchen, Klosterplatz 2, 4801 Traunkirchen, Austria. All rights reserved

References

1. Hans Joachim Specht *60 Jahre Physik - Faszination der Vielfalt.* https://books.ub.uni-heidelberg.de/heibooks/catalog/book/253/chapter/3935
2. Helmholtz, H. L. F. *Die Lehre von den Tonempfindungen als physiologische Grundlage für die Theorie der Musik* Braunschweig: F. Vieweg und Sohn (1863)
3. Seebeck, A. Beobachtungen über einige Bedingungen der Entstehung von Tönen. Annalen der Physik und Chemie, 53, 417–436 (1841)
4. Ohm, G. S. *Über die Definition des Tones, nebst daran geknüpfter Theorie der Sirene und ähnlicher tonbildender Vorrichtungen*; Annalen der Physik und Chemie, 59, 513–565 (1843)
5. Plomp, R. and Levelt, W.J.M. *Tonal Consonance and Critical Bandwith.* The Journal of the Acoustical Society of America, 38, 548–560 (1965)
6. Schouten, J. F. *The perception of subjective tones*; Proceedings of the Koninklijke Nederlandse Akademie van Wetenscbappen, 1938, 41, 1083–1093. *The residue and the mechanism of hearing*; Proceedings of the Koninklijke Nederlandse Akademie van Wetenschappen, 43, 991–999. *The residue: A new component in subjective sound analysis.* Proceedings of Koniklijke Nederlandsche Akademie van Wetenschappen, 43(3), 356–365

7. Peter Schneider, Michael Scherg, H. Günter Dosch, Hans J. Specht, Alexander Gutschalk and André Rupp *Morphology of Heschl's gyrus reflects enhanced activation in the auditory cortex of musicians* Nature Neuroscience volume 5, pages 688–694 (2002)
8. Peter Schneider, Vanessa Sluming, Neil Roberts, Michael Scherg, Rainer Goebel, Hans J Specht, H Günter Dosch, Stefan Bleeck, Christoph Stippich, André Rupp; *Structural and functional asymmetry of lateral Heschl's gyrus reflects pitch perception preference* Nature Neuroscience volume 8, pages 1241–1247 (2005)
9. N Sieroka, HG Dosch, HJ Specht, A Rupp—Neuroimage, Volume 20, Issue 3, November 2003, Pages 1697–1703 *Latency effect of the pitch response due to variations of frequency and spectral envelope,* Neuroimage, Volume 20, Issue 3, November 2003, Pages 1697–1703
10. S Ritter, HG Dosch, HJ Specht, P Schneider, A Rupp; *Latency effect of the pitch response due to variations of frequency and spectral envelope*; Clinical Neurophysiology Volume 118, (2007), Pages 2276–2281
11. S Ritter, HG Dosch, HJ Specht, A Rupp; *Neuromagnetic responses reflect the temporal pitch change of regular interval sounds*; NeuroImage Volume 27, Issue 3, September 2005, Pages 533–543
12. A. Rupp M. Hauck, HG Dosch, RD Patterson; *The Effect of Age on Huggins' Pitch Processing and its Location in Auditory Cortex* ACTA ACUSTICA UNITED WITH ACUSTICA Vol. 104 (2018) 783–786 DOI
13. Fan CS-D, Zhu X, Dosch HG, von Stutterheim C, Rupp A *Language related differences of the sustained response evoked by natural speech sounds* PLoS ONE 12(7): e0180441. https://doi.org/10.1371/journal.pone.0180441. (2017)
14. Colloqium at CERN 1998 https://cds.cern.ch/record/423897?ln=en

Exploring Psychophysics, Neuroscience, and Music

Peter Schneider

My collaboration with Professor Hans Joachim Specht and, in close partnership, Professor Hans Günter Dosch was rooted in a deeply interdisciplinary approach that seamlessly bridged physics, neuroscience, and music. Our work began in 1994 during their lecture series *"Helmholtz and Beyond,"* held to commemorate the 100th anniversary of Hermann von Helmholtz's death. The series delved into the physical principles underlying music and served as a catalyst for our research. Our discussions and experiments took place in diverse settings, from the *Große Hörsaal der Physik* to the homes of Professors Specht and Dosch in Heidelberg. These gatherings brought together an eclectic group of professional musicians and scholars, including Martin Messmer, a composer with absolute pitch and professor of music theory; Ulrich Charisius, a piano tuner and vice president of the German Piano Tuners Association;

and Doris Geller, a professor of ear training at the Mannheim-Heidelberg University of Music, alongside numerous invited musicians and scientists. These intimate sessions provided a unique platform for psychoacoustic experiments, focusing on the subjective perception of overtones (harmonics of complex sounds), fundamental tones, dissonance, and consonance. This collaborative and multidisciplinary environment greatly enhanced our understanding of the intricate interplay between sound, perception, and musical structure.

During these meetings, musicians were asked to use the available instruments—Professor Dosch's Blüthner grand piano, Professor Specht's Steinway grand piano—or their voices to express how they perceived the musical sounds being presented. Stimuli included mathematically calculated incomplete harmonic tones and digitally synthesized sounds, which were transmitted via headphones. These experiments sought to uncover interindividual differences in auditory analysis and listening preferences. Drawing inspiration from Helmholtz's groundbreaking work *On the Sensations of Tone as a Physiological Basis for the Theory of Music*, written in 1863 in Heidelberg, our discussions explored topics such as the perception of musical consonance and dissonance, the interplay between dispositional factors, musical aptitude, and expertise, as well as the nuances of long-term training in the context of the nature–nurture debate. To address these fundamental questions, we invited the American music psychologist Edwin Elias Gordon (1930–2015) and discussed with him the significance of his *Advanced Measures of Music Audiation (AMMA)* test, which assesses the perception of melodic and rhythmic differences. This provided an opportunity to examine the underlying concept of "audiation"—the mental representation of sound—and its essential role in musical imagination and perception.

These collaborative efforts culminated in my diploma thesis (1994–1996) on "residue and interval recognition in complex harmonic tones," supervised by Professors Specht and Dosch. Thanks to the outstanding expertise and dedication of Hans-Georg Siebig, the lecture assistant for the *Große Hörsaal der Physik*, we were able to utilize advanced technical equipment, including 40 infrared headphones from Sennheiser. This enabled a large-scale experiment investigating individual differences in the perception of timbre, fundamental tones, spectral aspects of harmonic complex sounds, musical intervals, and melodies. The study revealed that music perception, much like art, can manifest itself in various forms: a punctual perception of elements such as fundamental tones and residues, a linear perception of melodic progressions and harmonic structures, and a more surface-like perception of timbres and soundscapes. This conceptual framework aligns with the artistic principles of "point and line to plane" as articulated by Wassily Kandinsky (Fig. 4.33).

In 1997, our research expanded to investigate the neurological correlates of musical perception in collaboration with Professor Michael Scherg, a physicist and neuroscientist, and head of the Section of Biomagnetism at Heidelberg University Hospital. Professor Scherg was instrumental in bringing Germany's second magnetoencephalography (MEG) device to the Head Clinic in Heidelberg, enabling cutting-edge research in biomagnetism. Using this advanced MEG technology, we recorded brain activity in musicians, whose perception we had previously analyzed through psychoacoustic experiments, as they listened to harmonically complex tones and

Fig. 4.33 Hans Günter Dosch, Hans Joachim Specht, and Peter Schneider (from left to right), photographed immediately after Peter Schneider's doctoral examination in Heidelberg (Dean's Office, Albert-Überle-Weg, December 2000). © Peter Schneider. All rights reserved

musical sounds. Additionally, participants underwent MRI scans at the Neuroradiology Clinic in Heidelberg to study the structural characteristics of brain regions involved in sound and music processing, with a particular focus on the "Heschl's gyri," named after the Viennese anatomist Richard Ladislaus Heschl. These gyri, central to the auditory cortex, exhibited remarkable interindividual variability in shape, form, and size. Professor Specht's Heschl's gyri, for instance, were notably large and duplicated on both sides, a characteristic often observed in professional and highly talented pianists. It was therefore no surprise that during one of his lectures, with just a single day of preparation, he brilliantly and flawlessly performed the first movement of Frédéric Chopin's B minor Scherzo himself when no other pianist was available on short notice.

By integrating MEG and MRI, we analyzed both the electrical activity and structural characteristics underlying music perception, which formed the core focus of my doctoral research, titled *"Source Activity and Tonotopic Organization of the Auditory Cortex in Musicians and Non-Musicians."* The MEG study revealed that primary auditory cortex activity, occurring 19–30 ms after tone onset—representing fundamental processes of musical perception—was, on average, 87% greater in musicians compared to non-musicians, with professional musicians exhibiting the most pronounced effects. Late secondary activity, occurring around 100 ms after tone onset—more closely associated with selective attention processes—showed a smaller increase of 30%. Notably, the magnitude of primary source activity correlated strongly with musical aptitude, as measured by the AMMA test, though no correlation was observed with the age of musical training onset. These findings provide neurophysiological support for Gordon's (1987) hypothesis of a predominantly genetic predisposition for musical aptitude. The study also identified a robust logarithmic tonotopic organization in the primary auditory cortex, which intriguingly displayed the structure of a double mirrored "cortical keyboard." Lower frequencies were processed more superficially, at a resolution of 2.3 mm per octave on average. In contrast, late activity at 100 ms revealed a mirror-image tonotopic map in the adjacent planum temporale, where higher frequencies were processed more superficially, at a resolution of 1.45 mm per octave.

Our findings revealed significant structural and functional differences in Heschl's gyrus among professional musicians, amateur musicians, and non-musicians, strongly correlating with levels of musical training and aptitude. Furthermore, individuals exhibited distinct auditory processing strategies: some prioritized the fundamental frequency (f_0), while others focused on spectral components (f_{SP}). These preferences were linked to lateral asymmetries in Heschl's gyrus, such that spectral (overtone) listeners tended to have larger Heschl's gyri and stronger activation in the right auditory cortex, whereas fundamental frequency listeners showed larger gyri and stronger responses in the left auditory cortex. Notably, Professor Specht exhibited a balanced, mixed auditory perception profile, characterized by dominant fundamental frequency perception in the mid-frequency range and enhanced overtone perception in both higher and lower frequency ranges—a pattern typically observed in pianists. These findings, which integrated neuroanatomical, neurofunctional, and psychoacoustic aspects and led to the identification of specific individual "neuro-auditory sound perception profiles," were published in two collaborative *Nature Neuroscience* papers (Schneider et al., 2002 and 2005).

This pioneering work bridged physics, music psychology, and neuroscience, advancing our understanding of the neurophysiological basis of music perception. It also laid the foundation for public engagement, such as the Einstein Lectures Dahlem in 2006, where live experiments, theoretical insights, and musical demonstrations were combined. These interdisciplinary events underscored the transformative potential of our approach, offering profound insights into the intricate interplay between the physics of sound, psychological perception, and the neurological processes underlying musical understanding.

It Was Music Which Connected Us

Ulrich Charisius

It was music which connected us. For more than 40 years, it has accompanied Hans Joachim Specht and myself. But it was also my interest in physics that found a fatherly friend in him.

One day in the early 1980s—at that time, I was working as a master piano maker at Musikhaus Hochstein in Heidelberg—a gentleman entered the shop, introducing himself as Professor Specht. He had a special request: he planned to measure the acoustic characteristics of the Steinway concert grand piano in the local city hall (Stadthalle Heidelberg) and asked if someone would be so kind as to join him on this venture. This was exciting for me: capturing the sound data of a musical instrument alongside an experimental physicist, and what's more, an instrument that I regularly took care of in preparation for various concerts. Professor Specht wished to use the data for a lecture series at the University of Heidelberg's Studium Generale which focused on sound analysis of musical instruments and the neurophysiology of human sound perception. He was preparing this course together with his colleague and friend Hans Günter Dosch as well as other collaborators.

That was the beginning of our long-lasting friendship.

When the lectures started, I regularly attended. Sometimes I was asked to help with the preparation and follow-up of the experiments, and when needed, I tuned the old Bechstein piano for demonstration purposes before a lecture. Being a layman in physics, I did not understand everything presented scientifically, but nevertheless, from the very beginning, I felt that both the professors and the staff perceived me as an equal.

Professor Specht even invited me to explain the workings of a concert grand piano's action mechanism to his audience in one of these lectures. This was something very special for me: using a beamer image of the grand piano mechanism, I explained the complicated processes that occur mechanically while playing, and the audience was able to closely follow the movements of the individual parts of the mechanism.

Another encounter with Professor Specht allowed me to take part in the studies conducted by Dr. Peter Schneider. At the time, he was researching how sound reaches human consciousness. Sound perception ability was studied using magnetoencephalography (MEG) at Heidelberg Kopfklinik. In addition, hearing tests were conducted to determine which group of participants were fundamental tone listeners, which were harmonic listeners, and also whether there existed differences in sound perception between amateur and professional musicians.

All this research was initiated by Professor Specht and based on a shared appreciation of music in general and its subjective perception in particular.

Professor Specht was an excellent pianist. At some point in 1988, as I recall, a new piano was to be purchased for his home, a Steinway Model B-211. So, together, we drove to Hamburg and selected a particularly well-sounding specimen from several available B-pianos at the Steinway factory. For the entire Specht family, it was an outstanding experience to be present in this special room where all the most famous pianists, past and present, choose their individual instruments.

Of course, the new Steinway in the Specht household required regular tuning and maintenance. These appointments were always highlights of my work because Professor Specht and I used them as an opportunity to discuss special details of piano care. Professor Specht always took the time for these discussions. He was particularly interested in questions such as how the correct intonation of an instrument is achieved and what kind of influence inharmonicity has on the tuning process and the sound. In turn, he spoke to me about his research at CERN in Geneva and other projects he was involved in. He often played various pieces by Bach and Chopin from memory on the freshly tuned instrument.

In April 2015, Grigory Sokolov gave a piano recital at the Stadthalle Heidelberg. I was on concert duty and had prepared the Steinway Grand for the performance. During the intermission, I met Mr. and Mrs. Specht by chance, both of whom were thrilled by Grigory Sokolov's performance. They asked me if it would be possible to meet Mr. Sokolov in person, as they both greatly admired him. After the concert, I took the Spechts to the artist's backstage area of the Stadthalle, and they had the chance to actually meet him. Even years later, they both told me again and again how deeply touching this experience had been for them.

What I found remarkable was that, on the one hand, Professor Specht investigated the physics of musical instruments and the neurophysiological aspects of sound perception from a scientific perspective, while on the other hand, he enjoyed the music performed on these instruments from an artistic point of view. His preoccupation with music must certainly have given him a lot of strength and inspiration for his research work. In this respect, there was a connection between us: as a piano tuner, I hear sounds in ways defined by physics, but at the same time, I also have to do justice to the musical side of things. As I recently learned, there was yet another similarity we shared, which is that he had once actually wanted to become a sound engineer, just like myself.

Life—being born, growing old, falling ill and, eventually, having to pass on—then relentlessly took its course: while doing the groceries in Handschuhsheim one day in April 2023, I met Mrs. Specht, who informed me that her husband was having surgery on that same day, at that very hour. This news hit me like a bolt from the blue.

On 27 February 2024, I was scheduled to fine-tune the grand piano. I arrived at the apartment, saw Professor Specht, and was deeply shocked inside. However, I didn't let on, and we instead engaged in a factual discussion about the tuning issue. I had an idea and tuned a section in the lower treble range. He tried it, played for a bit, and said that the piano now sounded as beautiful as ever before. And in this exact moment, I felt that he had somehow come to terms with the inevitable for all of us.

That was my most intense and, at the same time, final encounter with Professor Specht.

I have tried to share my most extraordinary and memorable encounters with him. My connection with him, which had grown over more than 40 years and oscillated between music and physics, was shaped by both closeness and periods of distance. In the end, however, closeness and distance in such a relationship are one and the same, in a sense that totally defies the laws of physics, that is to say, they are ultimately and fundamentally human.

Hans J. Specht and the GSI Cancer Therapy Project

Hartmut Eickhoff

When Hans Joachim Specht became the Scientific Managing Director of GSI in autumn 1992, one of his priority projects was the treatment of cancer patients using carbon ions and the innovative raster scan irradiation method developed at GSI. Shortly after his appointment, he set his ambitious goal of achieving this milestone by the end of his tenure at GSI in 1999. Treating patients represented a complete shift from GSI's established focus of conducting only fundamental research.

Through his persuasive leadership and assertiveness, both in dealing with political authorities and overcoming internal resistance, Hans J. Specht successfully achieved his goal for this project. By May 1993, a collaboration was established between GSI, the University Clinic in Heidelberg, and the DKFZ. Thanks in particular to the preparatory work by Gerhard Kraft and Dieter Böhne, this collaboration resulted in the submission of a proposal to establish an experimental cancer therapy project.

Remarkably, less than 2 years after the proposal was approved by the GSI Supervisory Board, the construction of the treatment facility was completed. This included integrating the treatment area with the high-energy beam transport system and building additional infrastructure for patient preparation and facilities for medical and technical staff.

In contrast to these early construction efforts, the necessary modifications to the accelerator control system to meet the requirements of the raster scan irradiation modality were for a long time significantly underestimated. A clear implementation plan was not fully specified until 1995. This new treatment modality required rapid, reproducible, and reliable changes to multiple set values, for example, to vary the beam energy or the intensity of the ion beam on a time scale of seconds. This, in turn, was only achievable by using tested, stored data sets and taking into account the special safety requirements for patient irradiation.

Another challenge was that the requested beam parameters were dictated by the raster scan system, and their corresponding activation of the set values had to be carried out by personnel in the medical control room, rather than by the operators in the central accelerator control room, as in the usual experiments.

The control concept eventually developed and presented to the GSI management in 1995 was completely new, especially regarding accelerator operations, and its feasibility was initially uncertain. The resulting restrictions on experimental operations during treatment beam times in the coming years were another issue. Here too, Hans J. Specht played a decisive role, ensuring that these limitations were accepted almost without exception by the scientific community and that sufficient time would be allocated for the developments necessary for the therapy project.

With the strong motivation of the individuals involved in this project and the unwavering support of the GSI management, all technical challenges were successfully overcome, leading to the treatment of the first patient before the end of 1997 (Fig. 4.34).

4 Reflections and Tributes

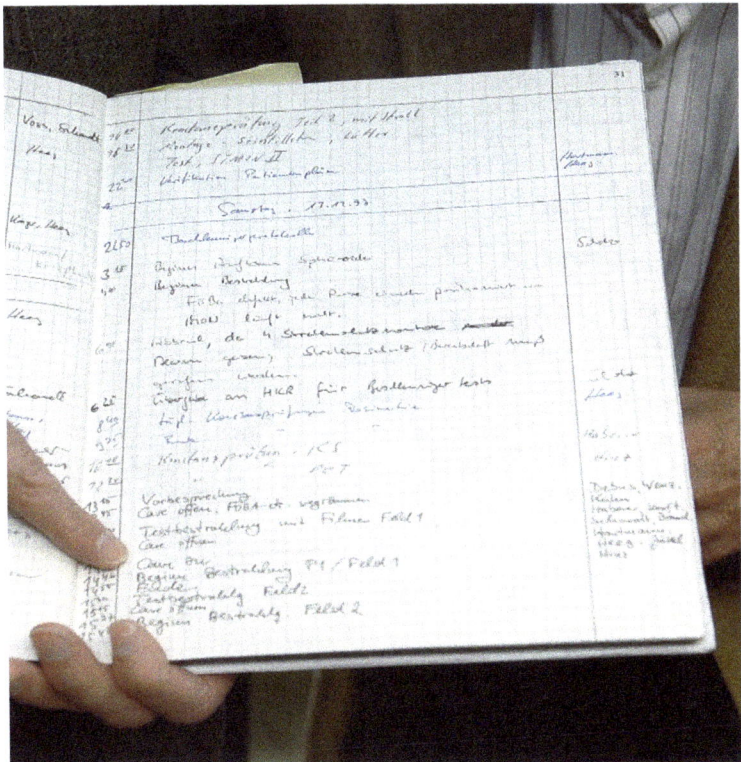

Fig. 4.34 Protocol book with notes for the first patient treatment on 13 December 1997. © FAIR-GSI. All rights reserved

Hans J. Specht expressed his deep passion for the GSI therapy project particularly impressively in his farewell speech as GSI Managing Scientific Director in 1999. In his speech, he reflected that his happiest moment at GSI was the successful completion of the first patient irradiation on 13 December 1997. Describing the atmosphere in the main control room during and especially after this groundbreaking treatment, he said: "The faces of those involved, the happiness in their eyes, the embrace shared among the team—this is unforgettable. That was the finest hour, and that's when we understood why we do science" (Fig. 4.35).

In the proposal from 1993, the GSI experimental therapy program was already defined as a precursor to a new clinic-based facility, the essential characteristics of which have already been outlined here. Even before patient treatment began, GSI and the Heidelberg Clinic worked together to expand this proposal for such a clinic-based therapy system and presented it to the university management in Heidelberg at the end of 1998.

In addition to the successful start to patient treatments at GSI, Hans J. Specht was able to initiate work on the HIT therapy facility at the Heidelberg University Hospital

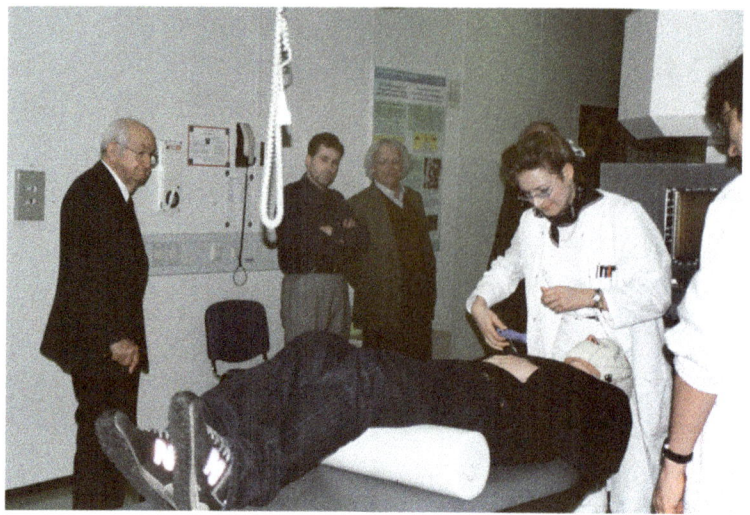

Fig. 4.35 Professor Schmelzer, Dr. Harberer and Professor H. J. Specht (from left to right) in the treatment room 'Cave M'. © FAIR-GSI. All rights reserved

during his tenure at GSI. Enthusiasm for this project led to the realization of the HIT facility, which applies light-ion irradiation to tumor patients. It began operation in 2009, following the termination of the GSI therapy project in 2008, after 440 patients had been successfully treated at GSI.

Advancing Inertial Fusion and Plasma Physics

Thomas Kühl and Andreas Tauschwitz

When Hans Specht started his term as Scientific Director of GSI in 1992, the plasma physics research group was a small unit, which had emerged a few years earlier and was primarily driven by the interest in inertial confinement fusion. This direction was initiated by Rudolf Bock, who had already advocated the idea at the DPG spring meeting in Heidelberg in 1978, and started the plasma physics activity within his department.

During his time as director of GSI, Hans Specht promoted plasma physics in many ways. Two important activities which he supported were the work on the Heavy-Ion-Driven Inertial Fusion (HIDIF) study and planning for a kilojoule-class laser facility to complement the unique possibilities of the heavy-ion beam, as a tool to create and investigate matter at extreme conditions of temperature and pressure.

Motivated by this research, the HIDIF study was conducted from 1995 to 1998 as a follow-up to the studies HIBALL I and HIBALL II, which were published in the years 1982 and 1985 and led by the Forschungszentrum Karlsruhe. In the

HIDIF study, GSI took the lead, playing on its strength in accelerator technology and experimental activities related to high-energy–density matter. Scientists from 16 institutions contributed to the study. Half of the authors were from GSI and the neighboring universities in Frankfurt and Darmstadt, demonstrating GSI's leading role in this field of research.

In his foreword to the HIDIF report, Hans Specht stated that "It should [ ] be the responsibility of the scientific community to advance the development of fusion energy and to explore without bias all possible ways to make this new energy source available as soon as possible. Inertial fusion has in the past decades in Europe been treated as a sideline to the main thrust oriented towards magnetically confined fusion. The complications and implications of the latter are, however, such that practical realization is still several decades away. A thorough investigation as to whether inertial fusion might, in the end, yield useful energy under more favourable conditions therefore appears to me a scientific necessity in present times." In the view of this study, the heavy-ion accelerator was the most promising driver for the production of electrical energy by Inertial Confinement Fusion (ICF). With the increased interest in ICF driven by recent achievements at the Livermore National Ignition Facility, this statement is still as relevant as it was more than 25 years ago. Various driver schemes for a possible future ICF power plant are currently under review and the HIDIF study is still a valuable source of information on the possibilities for a RF-driven accelerator for fusion applications.

Together with the HIDIF study, GSI engaged in a number of conferences, especially the 12th International Symposium on Heavy-Ion Inertial Fusion in Heidelberg, 24–27 September 1997 (Fig. 4.36). Through this international context, an idea was put forward to equip the facility at GSI with a powerful laser system, which would improve the feasibility of plasma research. Hans Specht was interested in this project from the beginning, and it was further developed under his guidance, in close consultation with the Laser directorate of the Lawrence Livermore National Laboratory in the USA, and with experts from leading laboratories in Europe and the USA. An international working group was set up, and in 1998 the resulting proposal for PHELIX the "Petawatt High Energy Laser for Heavy-Ion Experiments" was accepted and added to the "Long-Range Plan for an Upgrade of the GSI Facilities".

After completion, the PHELIX facility became part of the Laserlab Europe consortium as a user facility and serves the German and the international communities for cutting-edge research in plasma physics, but also atomic physics, biology, and materials science. Continued success was guaranteed by the addition of further experimental capabilities, including an X-ray laser laboratory, an experimental station within the PHELIX building, and beamlines towards the UNILAC experimental station Z6 and the HHT cave at SIS 18. These additions allowed for a full range of experiments with the combination of laser beams from PHELIX and heavy-ion beams from the accelerator. Scientists at GSI now have the unique opportunity to combine high-energy and high-intensity laser beams with ion beams, which are produced in the existing accelerator facility. Matter can thus be studied under extreme conditions, as they occur in stars or in the interiors of large planets, such as Jupiter. Given the need for better laser performance, novel ultrafast OPA technology, adaptive optics,

Fig. 4.36 HIDIF result presented at the 12th International Conference on Heavy-Ion Inertial Fusion, Heidelberg, September 1997. © FAIR-GSI. All rights reserved

and a number of other advancements have been implemented. An upgrade to higher repetition rate is currently under way.

PHELIX has been a highly successful program which has attracted a large German and international user community and greatly contributes to the attractiveness of FAIR/GSI research. Since user operation began in 2008, 164 experiments have been conducted in 177 experimental campaigns. The results of these experiments have been published in more than 200 papers in refereed journals. Beam time is in high demand and distributed selectively via an international advisory committee.

In accordance with Rudolf Bock and Hans Specht's vision, these capabilities are also used in experiments aiming to find a faster route to inertial fusion energy, in cooperation again with German and international researchers, both from academia and industry. The technological experience gathered within the project and the framework of GSI and FAIR will allow for the development of laser installations exploiting the new capabilities at FAIR.

Recollections of His Personal Assistant at GSI Darmstadt

Klaus Dieter Gross

Hans Specht took over the post of Scientific Director of GSI in the fall of 1992 and held this position until the fall of 1999. His term of office saw many outstanding scientific results at the GSI facilities as well as landmark decisions for the further development of GSI towards a European/international research center.

With regard to the governance of GSI, Hans Specht introduced a new management structure with a five-member executive board, which was based on the CERN model and was staffed with the following members, some of whom came from outside:

Chair and Scientific Director: Hans Specht;
Research Director: Volker Metag (University of Giessen),
from 1999 Jürgen Kluge;
Accelerator Director: Norbert Angert;
Scientific and Technical Infrastructure: Wolfgang v. Rüden (CERN);
Administration and Commercial Director: Helmut Zeitträger;
Secretary/Personal Assistant: Klaus-Dieter Groß.

In addition, Hans Specht pushed for further internationalization of the scientific advisory bodies, especially the Scientific Council, but also some expert committees, e.g., on biophysics and materials research, and consistently switched the language in these bodies from German to English.

Hans Specht had taken over the office from Paul Kienle, under whose aegis the new GSI facilities with the heavy-ion synchrotron SIS18, the fragment separator FRS, the experimental storage ring ESR, and numerous new detectors had been built and put into operation in the years 1984–1990. At the beginning of his term of office, the commissioning of the accelerators was largely completed and the new facilities were ready for full experimental operation.

Accordingly, there was a great spirit of enthusiasm in the community, and new exciting results tumbled in almost every beam time. The meetings of the Experimental Committee (EA)—as the GSI Program Advisory Board was then called—which met twice a year, were almost festive at GSI in those days. The presentation of new proposals was open to the public at that time and took place in a packed lecture hall, because everyone wanted to hear and see what new experimental ideas were in the pipeline.

This was an extremely euphoric and scientifically exciting time, and under the leadership of Hans Specht, it became possible to further optimize the yield of successfully completed cutting-edge experiments. To this end, he introduced new meeting formats, such as a round table on strategic beam time planning, which included not only the EA chairperson and beam time coordinator but also the leading GSI scientists.

The scientific reports 1992–1999 bear witness to the many scientific successes of that time. To name but a few (see also Fig. 4.37):

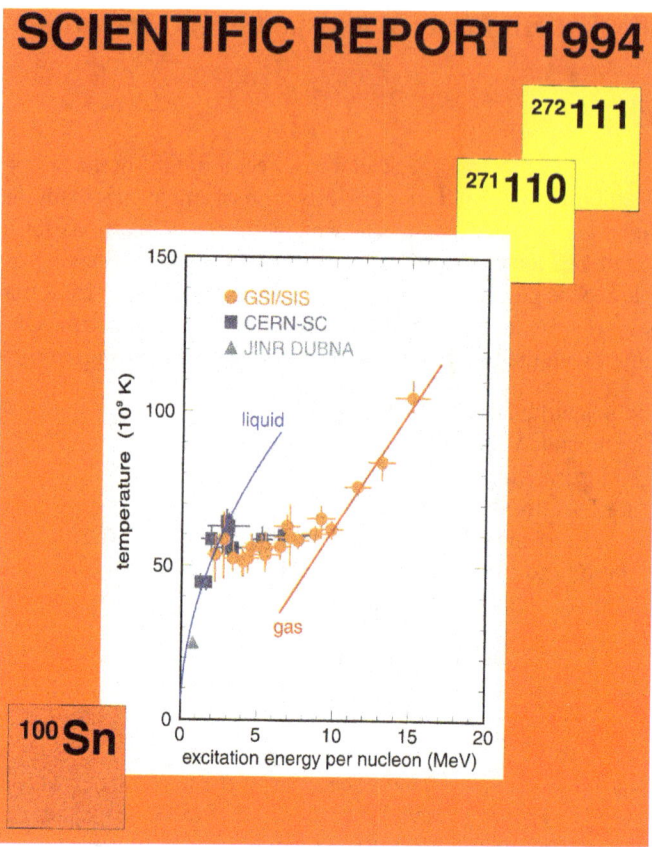

Fig. 4.37 Coinciding perfectly with GSI's 25th anniversary, 1994 was a particularly productive year for research, with three highlights decorating the front page of the Scientific Report. These were: (i) the production and identification of the long-sought doubly magic nucleus ^{100}Sn at the FRS; (ii) the first convincing demonstration of a liquid–gas phase transition of hot nuclear matter at the ALADIN spectrometer; and (iii) the discovery of two new elements with atomic numbers 110 and 111 by the SHIP group. © FAIR-GSI. All rights reserved

- Discovery of the chemical elements 110-darmstadtium, 111-roentgenium and 112-copernicium.
- Synthesis, identification, and study of many new exotic isotopes with extreme proton or neutron excess, including the doubly magic nuclei ^{100}Sn and ^{132}Sn, as well as ^{48}Ni and ^{78}Ni, at the FRS.
- Systematic mass measurements of many exotic nuclei using Schottky frequency spectroscopy or time-of-flight techniques at the FRS/ESR.
- Study of the bound beta decay of the cosmic clock nucleus $_{75}^{187}$Re^{75+} → $_{76}^{187}$Os^{75+} at the ESR.
- First observation of low-lying (1s) pionic states in Pb atoms at the FRS.

- Evidence for a liquid–gas phase transition in hot nuclear matter at the ALADIN spectrometer.
- Observation of the collective flow of light and medium-heavy fragments in highly central, high-energy Au + Au collisions at the FOPI detector.
- Evidence for medium modifications of kaons (K-mesons) in compressed nuclear matter at the KAOS spectrometer.
- Precision studies with highly charged ions, e.g., U^{91+} and Bi^{82+} ions, stored in the ESR, aiming at fundamental tests of QED in extremely strong electric and magnetic fields.

In the context of the scientific program, it is also worth mentioning—and typical of Hans Specht—that he pushed very decisively for a solution to the so-called e^+e^- puzzle at GSI. These were elusive line structures identified in the 1980s by the EPOS and ORANGE collaborations in the e^+e^- spectrum of heavy nuclear collision systems (e.g., U + Ta) near the Coulomb barrier. To throw light on this puzzle, long beam runs were allocated for a coordinated approach to the two experiments, which had since been upgraded. In these new high-statistics runs, the previously found signatures could not be reproduced.

In addition to the ongoing research, with the highlights listed above being just a small subset, important decisions on future research projects and collaborations were made during Hans Specht's term of office, some of which continue to have a positive impact until today. These include approval for the construction of the HADES detector at SIS18 and the HITRAP (Heavy-Ion TRAP) facility at ESR. In addition, two further projects were launched: the construction of the high-current injector at UNILAC to increase the intensities of heavy ions such as uranium by two to three orders of magnitude, and the realization of the high-power laser PHELIX (Petawatt High-Energy Laser for Heavy-Ion eXperiments) for the study of warm dense plasmas. Last but not least, the decision was taken that GSI would participate in the CERN LHC experiment ALICE, as an important part of the long-term cooperation between GSI and CERN.

Moreover, discussions about the long-term future of GSI were initiated during Hans Specht's term of office, with the significant involvement and coordination of the then Research Director Volker Metag, and with the broad participation of the European/international scientific community. The succeeding scientific director, Walter Henning, continued and expanded these discussions, which ultimately led to the proposal for the international FAIR facility (Facility for Antiproton and Ion Research).

The milestones and achievements during Hans Specht's time in office listed above already give an idea of both the scope and lasting impact of the scientific course and decisions taken in the years 1992–1999 as well as the wealth and relevance of the scientific results achieved at that time for the further development of GSI. And yet another milestone is missing, perhaps the most important one, which is capable of overshadowing all the others for its immediate social significance: **tumor therapy with ion beams**, which he energetically promoted and advanced during his time in office and beyond.

After Hans Specht had convinced himself of the medical relevance and technical feasibility of ion-beam therapy, he adopted this field as one of the focal points of GSI's scientific and technical program in 1993. The medium-term goal was to set up a pilot facility for tumor therapy for patient treatments at GSI, a decision that caused some controversy in the scientific community at that time and required both courage and determination.

In order to realize such a pilot facility at the GSI accelerator facilities, it was necessary to forge a powerful research alliance of institutions with scientific, technical, and above all medical expertise. Hans Specht played an essential role as both a guiding and a driving force in the establishment of this research cooperation, which, in addition to GSI, included the Department of RadioOncology and Radiotherapy at Heidelberg University Hospital (RadioOncology and Radiotherapy at UKHD), the German Cancer Research Center (DKFZ), Heidelberg, and the Helmholtz Center Dresden-Rossendorf (HZDR) as expert partner institutions.

At the GSI pilot facility, which went into medical operation in 1997 (see Fig. 4.38), around 450 patients, mainly with tumors at the base of the skull, were treated in clinical studies up to 2008. Subsequent observations over 5 years showed significantly improved therapy results and considerably reduced side effects compared to conventional radiation methods. Hans Specht himself commented on the importance of tumor therapy throughout his career: "There was hardly a happier hour in my professional life than after the first patient irradiation."

Based on the promising results of the pilot project, the cooperation partners GSI, RadioOncology and Radiotherapy at the UKHD and DKFZ started to work on a project proposal for the construction of a clinical therapy facility for tumor treatment with ion beams in Heidelberg.

I still remember this project proposal very vividly, as it was to be presented to the then Federal Minister of Research, Jürgen Rüttgers, during his visit to GSI in September 1998. The completion of the joint proposal was therefore under great time pressure. Everyone involved worked day and night, and we only managed to complete two bound copies of the proposal in the in-house print shop very late on the eve of the minister's visit. A precision landing, or so we thought, until Hans Specht picked one up and leafed through it as though it was a kind of flip-book. For the flip-book broke off halfway through the document and was followed by the exclamation "There's a mistake." After we had corrected the error (and a few others), two revised copies were bound the next morning and prepared for the ceremonial handover.

The minister's visit then went very well. Hans Specht presented the advantages and prospects of ion-beam tumor therapy in his usual brilliant manner—none of his colleagues from the medical side could have done it better. However, in the heat of the action he missed the moment of the ceremonial handover, and so the proposal later ended up somewhat unspectacularly in the Minister's car. But, as we all know, this did not detract from the successful progress of ion-beam therapy.

GSI also played a key role in the project for the Heidelberg Ion-Beam Therapy Facility (HIT), as it assumed responsibility for the construction and commissioning of the accelerator and the technical infrastructure for the HIT Medical Center. Construction began in 2004 and just 5 years later, the then Minister-President Günther

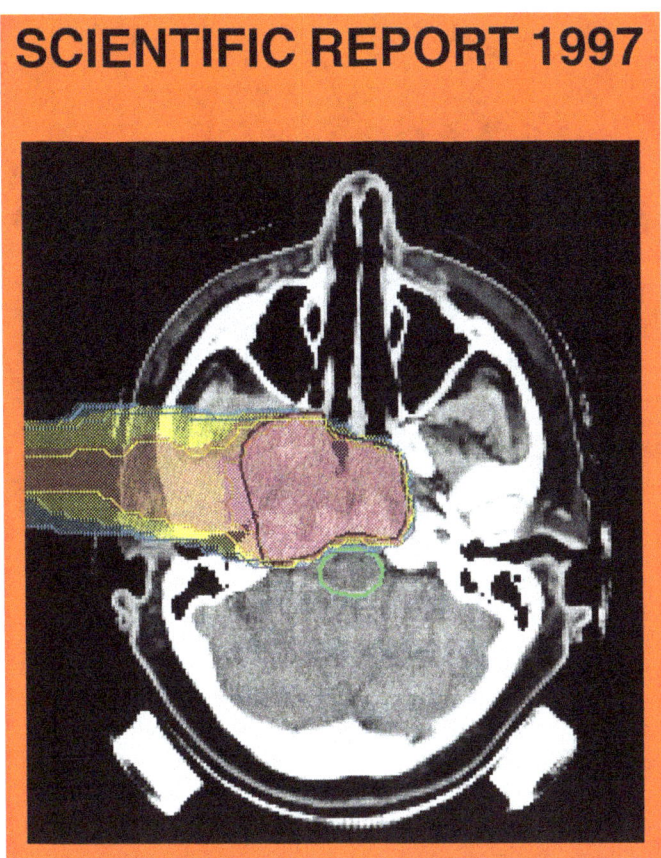

Fig. 4.38 The first patient treatments carried out using the new method of tumor therapy in 1997 belong without doubt to the most dramatic and significant moments in the history of the GSI research center. This is also reflected on the cover page of the corresponding scientific report. It shows the CT image of a patient overlaid with the enective dose distribution for single-field irradiation. The target area within the red line is very close to the brain stem (green circle), which must be spared. The colored areas correspond to 0–20% (blue), 20–40% (yellow), 40–60% (orange), and 80–100% (red) of the maximum dose. © FAIR-GSI. All rights reserved

Oettinger officially opened the therapy facility for patient treatments. Up to now (end of 2024), more than 9000 patients have benefited from ion-beam therapy at HIT. In 2015, a second clinical facility went into operation, the Marburg Ion-Beam Therapy Center (MIT) at the University Hospital Marburg-Gießen. The MIT was built by SIEMENS AG based on licenses from GSI and put into operation under the technical and medical management of the HIT. Since 2019, the MIT has been operated solely by Marburg University Medical Center. More than 2850 patients have been treated there.

Fig. 4.39 Raising a glass on the occasion of Hans Specht's 59th birthday in June 1995. From left to right: Klaus Dieter Groß, Jürgen Kluge and Hans Specht. © FAIR-GSI. All rights reserved

In retrospect, GSI experienced remarkable prosperity in research in the years 1992–1999, mainly, of course, due to the new GSI accelerator and experimental facilities that went into operation in 1990, but also, in my personal memory and opinion, thanks to the very clear, determined and focused scientific direction and leadership of Hans Specht. During this period, important strategic decisions were taken that still have a major and positive impact today, both in terms of basic research at GSI, including scientific cooperation with other leading research institutions, and also in terms of specific applications with large social benefits, in particular the tumor therapy with ion beams. One can only speculate about the role and importance that these strategic decisions and their successful implementation played in the German government's approval of GSI's long-term plans with the international FAIR project.

GSI looks back on Hans Specht's service as Scientific Director of GSI with the greatest respect and gratitude, and I personally associate many positive and lasting memories with this time and the occasional contact with Hans Specht thereafter (Fig. 4.39).

Twenty-Five Transformative Years with Hans J. Specht

Sanja Damjanovic

It is a rare privilege to work closely with a scientific luminary for an extended period of time. For me, that privilege spanned 25 transformative years under the mentorship of Hans J. Specht. Our professional path traversed three major chapters: the CERES experiment during my Ph.D. in Heidelberg, the NA60 experiment at CERN, and the establishment of Europe's fifth Hadron Cancer Therapy facility—the SEEIIST project—during my tenure as Minister of Science in Montenegro. In each of these chapters, Hans's mentorship played a pivotal role, shaping not only my scientific work but also my approach to science-policy development.

CERES and Mentorship: My Heidelberg Ph.D. Years with Hans J. Specht

Our professional path together began in the autumn of 1999 when Hans returned to the Physikalisches Institut in Heidelberg after his tenure as the Scientific Managing Director of GSI. At that time, I was one year into my Ph.D. in Heidelberg, tasked with studying thermal dileptons produced in Pb-Au collisions at the lower SPS energy of 40 AGeV. This was the first low-energy run for such a study, aimed at exploring the QCD phase diagram under conditions of higher baryon densities. It was also the first CERES run with an upgraded experimental setup. The original CERES detector system, Hans's "brainchild" affectionately known as the "Spechtometer," had been upgraded with a Time Projection Chamber (TPC). However, the 1999 run with the upgraded TPC faced significant setbacks, resulting in a TPC pair efficiency of only 19%. This created critical obstacles for dilepton analysis, as the magnetic field between the two RICH detectors had been switched off, leaving the TPC to function as the mass spectrometer. This challenging situation required a miracle, a combination of ingenuity and perseverance to salvage meaningful results.

The Power of Hans's Mentorship: Turning Setbacks into Results

The situation seemed almost hopeless for dilepton analysis. Dilepton pairs are already rare, with production probabilities orders of magnitude lower than those for hadrons, and separating them from both physical and large combinatorial backgrounds presented significant difficulties. Additionally, I faced the challenge of transitioning from my theoretical background in physics to becoming an experimentalist—a process that was anything but easy.

Then came a pivotal moment: Hans returned from GSI and asked me whether I would like to become his Ph.D. student. **That question marked the happiest moment of my professional life**. It was the beginning of a mentorship with Hans that turned each day into a learning experience. He not only deepened my understanding of experimental science but also instilled in me the confidence to navigate challenges. He encouraged me to argue with him—a dynamic that lasted 25 years and profoundly shaped my thinking. His mentorship was both rigorous and holistic, shaping my work in physics while also broadening my perspective on cultural topics. Hans's diverse interests—from music, cars, and photography to technology, electronics, engineering, and watches—were always intertwined with his deep understanding of physics, making them even more inspiring.

Hans's special ability to tackle setbacks with both confidence and joy—widely recognized and often spoken about—shone through once again as he addressed the challenges of dilepton analysis with such determination that any doubts about achieving meaningful dilepton results quickly dissipated. Working alongside postdoc Kiril Filimonov, I witnessed Hans's relentless drive firsthand. His methodical approach was admirable, often involving fast and accurate paper-and-pencil calculations that outpaced computational methods. Hans always carried a small, sharp pencil and an eraser in his pocket, symbols of his meticulous problem-solving approach. His daily guidance not only moved the analysis but also helped me become a confident experimentalist.

Then, the miracle we'd been waiting for happened. The dilepton pairs from 40 AGeV Pb-Au collisions appeared in the invariant-mass spectrum, showing an excess five times higher than the expected yield. The results far exceeded the values observed at the higher beam energy of 160 AGeV and were a significant step forward. Indeed, they saved my Ph.D.

The results were first presented at the 4th International Conference on Physics and Astrophysics of Quark–Gluon Plasma (ICPAQGP) in India in 2001. True to character, Hans ensured that I, as the Ph.D. student, had the opportunity to present these results—my first talk at a major international conference.

This success led to the publication of my final Ph.D. results in *Physics Review Letters (91, 2003, 042301)*, a milestone paper that remains the only study to date of dileptons at lower beam energies. The data were interpreted as direct radiation from the fireball, dominated by pion annihilation with a modified ρ-spectral function. While the exact mechanism of chiral symmetry restoration—mass shift versus broadening—remained unresolved, the larger enhancement factor at lower beam energies underscored the critical role of baryon densities in these processes.

The Lasting Impact of a Great Mentor

This chapter of my early career is a testament to Hans's ability to turn obstacles into meaningful results—a rare quality that only those who had the privilege of working with Hans can truly understand. His mastery in guiding complex analysis, steadfast determination, and commitment to his students' success, while treating them as equal colleagues, defined his mentorship. Hans upheld excellence with rigorous standards, but in a way that fostered growth in each student, enabling them to rise to their potential without feeling overwhelmed or discouraged. He entrusted each student with significant responsibilities and was always there to guide and support them, demonstrating unwavering trust in their abilities. Hans's influence extended far beyond research; he fostered critical thinking, resilience, and the ability to find joy in facing challenges. His guidance shows how a great mentor can have a lasting and transformative impact, both on individuals and on the broader scientific community.

While CERES marked the beginning of my professional path with Hans, the next chapter took us to CERN's NA60 experiment, where our work further deepened the understanding of the QCD phase transition.

NA60 at CERN: My Continued Journey with Hans at CERN's NA60 Experiment

Building on Hans's mentorship, I had the privilege of continuing to work with him as his postdoc on the next-generation heavy-ion NA60 experiment at CERN. This marked the beginning of an exhilarating and highly productive new chapter in my career. Under Hans's guidance, our work led to a series of groundbreaking results that revealed the thermal origin of dileptons in nuclear collisions at SPS energies. These findings provided compelling proof of the quark–gluon plasma, offering clear evidence of deconfinement and profound insights into the mechanism of chiral symmetry restoration—two cornerstone phenomena of QCD. Together, these results addressed long-standing questions central to CERN's ultra-relativistic heavy-ion program since its inception in 1984.

By taking a significant step forward in technology, the NA60 experiment set a new benchmark for dilepton data quality. It achieved an unprecedented sample size of 360,000 dimuons, with effective statistics up to three orders of magnitude higher than any previous results. Its improved mass resolution of approximately 2% enabled us to distinctly resolve all resonances, including the $\eta \rightarrow \mu^+\mu^-$ peak. As an upgrade to the earlier NA50 experiment, which focused on J/ψ measurements, NA60 opened new possibilities for studying low-mass dimuons for the first time.

Hans's Guidance: Setting New Standards in Dilepton Analysis

Hans's decision to join NA60 (after completion of all other SPS experiments, including CERES) greatly benefited the field. His expertise addressed a gap in low-mass dimuon analysis and set new standards for this area of research. As his postdoc, I had the privilege of working closely with him once again, sharing responsibilities for low-mass dimuon analysis.

This period was marked by my steepest learning curve and the immense joy of working with Hans within CERN's inspiring international environment. It was Hans's third sabbatical at CERN, following his work on the development of HELIOS and CERES. His deep insights and mastery of dileptons were immediately apparent. Hans quickly recognized that the exceptional data quality allowed us to isolate the dilepton excess using entirely new methods, applying local criteria and relying solely on the measured mass distribution, without any prior assumptions about its nature or model-dependent fits (Fig. 4.40).

The Breakthrough: First Measurement of the ρ Spectral Function

Our first breakthrough came with the measurement of the space–time averaged ρ-spectral function. The results, independent of theoretical modeling, demonstrated that "in-medium" effects manifest themselves solely as broadening—described as the "melting" of the ρ meson, without any shift in its mass. This groundbreaking result conclusively resolved a decades-long controversy about the spectral properties of hadrons near the QCD phase boundary, ending over 20 years of debate.

The moment we isolated the ρ spectral function and it appeared on our computer screens remains one of the happiest moments of my professional life—a joy shared with Hans and one that words can scarcely capture. Presenting this result publicly for the first time at the ECT* Trento Electromagnetic Workshop in May 2005, in front of Gerald (Gerry) Brown (co-originator of the mass-dropping hypothesis with Mannque Rho) and Ralf Rapp (a proponent of the broadening scenario with Jochen Wambach), was equally unforgettable. This result became a highlight of the Quark Matter 2006 conference in Budapest and was published in *Physical Review Letters 96 (2006) 162302*.

Fig. 4.40 Working with Hans in CERN's inspiring international environment. © Sanja Damjanovic. All rights reserved

Once again, Hans showed one of the hallmarks of great mentorship—always fostering the growth of younger team members. True to his tradition, he sent me up front for the first presentation of these results, including the delivery of a plenary talk at QM2006, attended by more than 1000 participants. I recall Hans's words: *"Throughout my time as a professor, mentor, and spokesperson, I have always put my young team members in front."* During this period, numerous invitations for seminars and colloquia were received, often with me as the featured speaker and Hans in the audience—a testament to his greatness. His voice was always one of the strongest at conferences, offering critical guidance and asking profound questions, it was always a pleasure to listen to him. Thanks to Hans's mentorship, my confidence in presenting results and engaging in critical discussions grew. It was also during this period that I was awarded a fellowship position at CERN.

Hans's dedication and passion for research were unparalleled—we often worked seven days a week at CERN, including weekends. Yet, even during the most intense periods, Hans knew how to balance the demands of rigorous research with what could be called active breaks, whether it was a quick swim in Lake Geneva, skiing in the nearby Alps, or attending music concerts at venues like the Verbier music festival or Victoria Hall, or sailing on the lake. Hans held the CERN environment as a special place (Fig. 4.41).

Though I come from Montenegro, my first sailing experience was on Lake Geneva through CERN's Yachting Club (YCC). In 2010, a lottery was organized for new sailing students, and it was Hans and Rob Veenhof, both passionate sailing enthusiasts, who encouraged me to apply for the dinghy course. At the time, I was completely unfamiliar with sailing and had no idea what a dinghy or an Yngling even was.

Both Hans and Rob, independently, assured me that my personality would be a perfect fit for this sport. Buoyed by their confidence, I entered the lottery on the

Fig. 4.41 An active break—skiing in Flaine—while writing a paper on eta and omega Dalitz form factors, the first high-energy heavy-ion results included in the Particle Data Group (PDG) compilation. © Sanja Damjanovic. All rights reserved

very last day and was fortunate enough to be selected for the dinghy course. I was even luckier to have Andrea Messina as my coach—the best club had—thanks to Rob's behind-the-scenes efforts. That first experience with sailing turned me into an instant enthusiast. Just one year later, I bought my own Laser sailboat—and was sailing 150 days a year, even during winter when the lake was iced over and all the YCC dinghies were removed from the harbor.

Hans and Rob were always there to guide and protect me from disaster. On one memorable occasion, Hans even organized a rescue mission when I found myself caught in stormy weather, capsizing every few seconds. Hans and Rob managed to pass their passion for sailing on to me. What started with frequent capsizing soon transformed into winning regattas and even being coached in groups by Olympic trainers.

None of this would have been possible without Hans and Rob's encouragement. Their wisdom and support opened the door to an incredible new journey for me—one that changed my life in ways I could never have imagined.

A Notable Recognition of Hans's Efforts

In 2003, despite his official retirement in 2004—though Hans never truly retired—he received federal funding from the BMBF (lasting until 2006) to contribute to NA60, supporting hardware development and postdocs. In 2006, the BMBF established a special evaluation committee to review all grants, and the results of NA60 were prominently featured on the top cover of the Report of the BMBF Review Committee "Hadrons and Nuclear Physics" 2003–2006, an honor that underscored the profound impact of Hans's contributions and the respect he commanded from the BMBF.

A Dream Realized: The Planck-Like Thermal Spectrum

Hans's critical insight guided the next groundbreaking result—the first measurement of a true thermal spectrum. A 25-year-old dream had become a reality with the measurement of a Planck-like thermal radiation spectrum, a prominent signal of quark–gluon plasma formation. Hans's meticulous attention to every aspect of the analysis made this milestone possible.

By leveraging NA60's advanced silicon vertex tracker, for the first time in nuclear collisions, prompt dimuons could be disentangled from offset pairs originating from D-meson decays. This analysis, largely carried out by Ruben Shahoyan, enabled the isolation of the excess also for masses above 1 GeV, now by subtracting the measured known sources: Drell-Yan and open charm.

Hans emphasized that, due to the Lorentz invariance of mass (M), the spectrum is unaffected by the expansion of the fireball, meaning that the temperature (T) derived from the spectrum reflects the purely thermal state of the system. By fitting the data with a Lorentz-invariant "Planck-like" function above 1 GeV, we derived a thermal temperature of 215 ± 12 MeV, well above the critical temperature of 160–170 MeV for the QCD phase transition, confirming partonic sources as the dominant origin of thermal radiation.

The Rise and Fall of the Radial Flow of Thermal Dimuons

Hans's guidance also extended to the analysis of transverse momentum (p_T) spectra. He championed the advantage of dileptons, which, unlike real photons, are characterized by two variables: mass and transverse momentum p_T. Quite different from mass, p_T not only contains contributions from the spectral functions, but also encodes the key properties of the expanding fireball: temperature and transverse expansion ("radial flow"). The final p_T spectra keep a memory of the time-ordering of the different dilepton sources produced from the fireball, offering a diagnostic tool for the emission region. By analyzing the transverse mass spectra ($m_T = \sqrt{(p_T^2 + M^2)}$), which are found to be nearly exponential, we could distill the data into a single effective temperature (T_{eff}), combining thermal contributions and radial flow effects ($T_{\text{eff}} = T + Mv^2$).

A crowning achievement followed with the measurement of the mass dependence of the effective temperature, revealing the "rise and fall of radial flow" of thermal dimuons—a memorable phrase invented by Hans. This clearly distinguished two sources: hadronic decay radiation with a "rise" in the radial flow up to 1 GeV, and partonic radiation signaled through a sudden "fall" in the radial flow at 1 GeV, followed by a nearly constant T_{eff} of about 200 MeV. This provided direct evidence for thermal radiation of partonic origin in nuclear collisions. These landmark findings were published in *Physics Review Letters 100 (2008) 022302* and *Eur. Physics J C 59 (2009) 607*, and summarized in *CERN Courier 49N9 (2009) 31–34*.

Contributions to Hadron Physics

Further groundbreaking results, coordinated by Hans, but now in hadron physics, were the high-precision measurements of electromagnetic transition form factors of η and ω Dalitz decays ($\eta \rightarrow \mu^+\mu^-\gamma$ and $\omega \rightarrow \mu^+\mu^-\pi^0$), which were published in *Physics Letter B 677 (2009) 260*. These results were included in 2010 in the *Particle Data Groups (PDG)* compilation, making them the first ever results from a heavy-ion experiment in the PDG.

One moment during this analysis remains vivid in my memory—the first time I showed Hans the preliminary results of the isolated Dalitz decays and the corresponding form factors. Hans immediately said, "There is a bug." This was just one of many such instances, and I always admired—though never fully understood—how he could instantly identify these issues. If you argued with Hans, he would quickly grab his pencil and prove the issue, always with a smile. The speed and ability with which he calculated rates were astonishing. His approach to verification—double-checking the results manually—was a testament to his deep understanding of physics.

Hans's Legacy: A Lasting Impact on Dileptons and Heavy-Ion Collisions at SPS

The Planck-like thermal dimuon mass spectrum, with the direct measurement of temperatures, presents the only solid proof of deconfinement at SPS energies. The study of dileptons in heavy-ion collisions at SPS and Hans J. Specht—a long journey together—culminated in great satisfaction and joy for Hans. Nearly two decades later, the enduring quality and relevance of these measurements, crafted under Hans's meticulous guidance, remain unsurpassed.

Working with Hans: Always Racing for the Next Groundbreaking Results

Working with Hans was like being part of an unending race towards new results—an exhilarating journey where each day brought new knowledge. His relentless drive for scientific excellence transformed every moment at CERN into an inspiring challenge. Hans was not satisfied with results that were merely good enough—he always sought the next steps, the next breakthrough.

SEEIIST: Hans's Mentorship in Science-Policy and the Launch of the First Hadron Cancer Therapy Facility for South East Europe

The race for impactful results—now with strong benefits for society—continued during my tenure as the Minister of Science of Montenegro, with Hans as my official and principal advisor and steadfast strategic supporter. On the very first day I assumed office, Hans's memorable words were: "That is great, but what are you going to work on tomorrow?" With his characteristic directness, Hans challenged me to think beyond the present and focus on what could create a lasting impact. This pivotal question became the catalyst for initiating the transformative SEEIIST project: the establishment of a large-scale Research Infrastructure for the Western Balkans and wider South East Europe (South East European International Institute for Sustainable Technologies). SEEIIST aimed to achieve what many deemed impossible: to bring cutting-edge technologies to a region that had gone through over seven decades without developing any large-scale research infrastructure.

For most, this mission seemed insurmountable. But not for Hans. He instantly encouraged the idea with boundless enthusiasm. Drawing on his vast experience in science management and his unparalleled international network, Hans started connecting me with influential figures from the global scientific community. Together, we moved the project forward, evolving it into a flagship initiative that bridged nations, united the international scientific community, and intertwined science, diplomacy, and societal progress in ways that exceeded all expectations.

For me, SEEIIST was far more than just a project; it was a testament to the 17 years of mentorship under Hans. His guidance instilled in me the courage, resilience, and ethical leadership needed to navigate such a complex and ambitious project—and to do so with joy.

One might well wonder how such an achievement was possible.

Hans's Role in Shaping SEEIIST as a Hadron Cancer Therapy and Biomedical Research Facility

At its core, SEEIIST was conceived as a platform for collaboration in a historically fragmented region, with the mission of "Science Diplomacy" or "Science for Peace." However, a question arose: What elements should SEEIIST contain in order to have a direct impact on society and easily attract crucial political support in the region, in particular, from the European Commission?

The proposal for SEEIIST to become an *Accelerator-Based Research Infrastructure for Cancer Therapy and Biomedical Research with Ion Beams* came in early 2017 from Hans, the pioneer of ultra-relativistic heavy-ion physics at CERN and hadron cancer therapy during his tenure as Scientific Director of GSI, where the first 450 patients in Europe were treated using this innovative method. Together with Hans and Nicholas Sammut, we spent over 100 h rigorously evaluating potential infrastructures to ensure the pan-European dimension of SEEIIST before reaching the final proposal in March 2017.

A key contributor to SEEIIST's Concept Study—the first phase of its development—was Ugo Amaldi (from CERN and the TERA Foundation), another pioneer in the design of particle accelerators for cancer treatment. Ugo played a pivotal role in establishing Europe's second hadron cancer therapy clinic, CNAO, in Pavia, Italy.

Hans's role in placing hadron cancer therapy and biomedical research Infrastructure at the core of SEEIIST, complemented by Ugo's role in developing the Concept Study, was instrumental in attracting political support from 10 South East European (SEE) countries, but also support from the broader European research and hadron cancer therapy community, including CERN, GSI, and the four existing European hadron cancer therapy clinics (HIT, CNAO, MIT, and MedAustron).

Under the guidance of these two pioneers in Hadron cancer therapy and as the first chair of the intergovernmental SEEIIST Steering Committee, I felt an unwavering confidence that every next step in the development of this complex project would succeed. Their guidance, wisdom, and our collective passion brought immense joy as we reached milestone after milestone.

One of the Milestones—The First Financial Support from the European Commission

One of the pivotal milestones in SEEIIST's journey was the first consideration of financial support from the European Commission, a moment that occurred during a Forum on New International Research Facilities for South East Europe organized by the Ministry of Science of Montenegro and the International Centre for Theoretical Physics (ICTP) in Trieste in January 2018. This Forum marked the first public presentation of SEEIIST, which was given by me. The event was attended by representatives from the European Commission (EC), led by Robert-Jan Smits, Director General of the DG-RTD, as well as representatives of the International Atomic Energy Agency (IAEA), prominent representatives of international and European research centers, including key figures in the field of particle accelerators and hadron cancer therapy, along with the regional scientific and medical communities (Figs. 4.42 and 4.43).

A particularly significant turning point was a presentation by Jürgen Debus, Medical Director of the first Hadron cancer therapy clinic in Europe, HIT in Heidelberg. That session, chaired by Hans with Jürgen on stage, was deeply emotional and made a strong impact on Robert-Jan Smits. Jürgen, who had been a former student

Fig. 4.42 SEEIIST Forum at ICTP Trieste in January 2018. This forum marked a significant milestone for SEEIIST. Left panel: Sanja Damjanovic (Minister of Science of Montenegro), Hans J. Specht, and Robert Jan Smits (DG of DG-RTD, European Commission) immediately following Sanja Damjanovic' first public presentation of SEEIIST. Right panel: Robert Jan Smits, Sanja Damjanovic, Hans J. Specht, and other Forum participants. © GOV.ME, photograph: Sasa Matic. All rights reserved

Fig. 4.43 Visit to the large-scale research infrastructure ELETTRA, Trieste, following the end of the SEEIIST Forum in Trieste. Left panel: Hans J. Specht with Jaques Balosso in the background. Right panel: Sanja Damjanovic with Hans J. Specht and other Forum participants. © GOV.ME, photograph: Sasa Matic. All rights reserved

of Hans in Heidelberg and the lead medical practitioner for the first patients treated at GSI under Hans's directorship, spoke passionately about his enduring respect for his mentor. Despite his demanding schedule treating patients, Jürgen told how he could never say no to his great professor, Hans J. Specht, and would never refuse to stand by him.

Their heartfelt connection, coupled with Jürgen's data-driven presentation of patient outcomes from HIT, and the visibly strong support from the political and international scientific community left a lasting impression on Robert-Jan Smits. The result was a breakthrough: the European Commission's initial consideration of financial support for SEEIIST.

Accelerating Progress

In the following months, the European Commission formally approved an initial 1 million EUR to start SEEIIST's second development phase—the Design Phase. This milestone was consolidated during my meeting with Carlos Moedas, the European Commissioner for Research at the time. Hans's strategic thinking was again important, and Jürgen once again demonstrated his unwavering dedication to Hans by traveling from Heidelberg to Brussels and back in a single day to join my meeting with Commissioner Moedas. As a result, we secured the first critical funding from the EC, enabling SEEIIST to move forward.

The Design Phase officially commenced with a high-level Kick-off event in Budva, Montenegro, in September 2019, attended by leading political and scientific figures, including representatives from the European Commission. From there, the momentum accelerated. Together with Hans, we worked on every single approach, whether it was securing regional political support, gaining backing from the European Commission, or garnering support from the international scientific community, all of which created a strong dynamic. We also prepared an IAEA application for capacity building and gifted it to the Clinical Center of Montenegro for them to submit. This resulted in funding of 500,000 EUR (Fig. 4.44).

Hans became the Chair of the SEEIIST Site Selection Committee, while Ugo continued developing the Technical Design Report, an essential milestone for the second development phase—the Design Study.

Hans was a strong advocate of international cooperation, which is the backbone of SEEIIST. The project has received robust international support from leading research centers and clinics from 14 European countries. This collaboration was further strengthened through additional European Commission funding under the EU H2020 HITRIplus project, coordinated by Sandro Rossi, Director of CNAO, and co-coordinated by Maurizio Vretenar from CERN. HITRIplus is paving the way to finalizing SEEIIST's design phase.

Today, SEEIIST is more than just a project; it stands as a symbol of progress in developing next-generation ion therapy facilities and expanding access to life-saving treatments for a much larger number of patients. For me, SEEIIST is also a deeply personal legacy of mentorship—a testament to the transformative guidance of Hans and his belief in the power of science to benefit society, as well as the true approach to science diplomacy. SEEIIST has gained significant recognition in EU agendas,

Fig. 4.44 High-level SEEIIST Kick-Off Event in Budva, Montenegro in September 2019, marking the official launch of the SEEIIST Design Phase, featuring a special discussion with the Prime Minister of Montenegro, Duško Marković. Left Panel: Prime Minister Duško Marković and Hans J. Specht, with Sanja Damjanovic and Aleksan Roy in the background. Right Panel: Sanja Damjanovic and Hans J. Specht alongside high-level representatives from the European Commission (EC), COST, ICTP, Central European Initiative (CEI), the Swiss Government, GSI, and CERN, during discussions with the Prime Minister of Montenegro. © GOV.ME, photograph: B. Čupić. All rights reserved

becoming part of the Economic Investment Plan for the Western Balkans (part of the Global Gateway initiative) and recognized as one of the six leading EU-WB projects in research and innovation in the Western Balkans' innovation agenda.

If SEEIIST reaches its full realization, it should be dedicated to Hans J. Specht and Ugo Amaldi.

In April 2024, during our last period of work together at CERN, Hans radiated his characteristic optimism and positivity, full of plans for the next steps. Even as his time came to a close, he filled the main CERN auditorium with music, joyfully playing the piano, performing Chopin and Schumann with the same enthusiasm he always radiated (Fig. 4.45).

Hans's final words to me continue to guide me: "*Never give up.*"

Fig. 4.45 Hans J. Specht at CERN, April 2024, in his last month, full of optimism and plans for the next steps. © Sanja Damjanovic. All rights reserved

Open Access This chapter is licensed under the terms of the Creative Commons Attribution 4.0 International License (http://creativecommons.org/licenses/by/4.0/), which permits use, sharing, adaptation, distribution and reproduction in any medium or format, as long as you give appropriate credit to the original author(s) and the source, provide a link to the Creative Commons license and indicate if changes were made.

The images or other third party material in this chapter are included in the chapter's Creative Commons license, unless indicated otherwise in a credit line to the material. If material is not included in the chapter's Creative Commons license and your intended use is not permitted by statutory regulation or exceeds the permitted use, you will need to obtain permission directly from the copyright holder.

Chapter 5
Closing Thoughts

Sanja Damjanovic, Volker Metag, and Jurgen Schukraft

As we bring this memoir and biography of Hans J. Specht to a close, we celebrate not only a renowned scientist whose career spanned nearly seven decades but also an extraordinary individual who inspired generations through his dedication, humility, and passion for science.

This book was born from a shared desire to honor Hans J. Specht's legacy, to document the depth and breadth of his contributions to physics and society, and to capture the essence of his life—both professional and personal. The resonance we encountered in the scientific community while assembling this book speaks for itself. As co-editors, we are deeply grateful to all the contributors who made this possible—friends, former students, colleagues, and collaborators whose insights and memories have enriched this tribute. These contributions, written in Europe, Japan, the United States, Canada, India, Australia, and beyond, reflect not only Hans J. Specht's global influence but also the lasting imprint he left on the scientific community across generations and continents.

Hans often spoke of the profound influence of his mentors, who guided him on his path, and this book stands as a testament to the enduring connections between generations of scientists. It reminds us how inspiration and knowledge are passed on, shaping not only the future of science but also the character of those who pursue it.

S. Damjanovic (✉)
GSI Helmholtz Centre for Heavy Ion Research, Darmstadt, Germany
e-mail: sdamjano@cern.ch

V. Metag
II. Physikalisches Institut, Justus-Liebig-Universität Giessen, Giessen, Germany
e-mail: volker.metag@exp2.physik.uni-giessen.de

J. Schukraft
CERN, Geneva, Switzerland
e-mail: jurgen.schukraft@cern.ch

© The Author(s) 2025
H. J. Specht et al. (eds.), *Hans Joachim Specht*, Springer Biographies,
https://doi.org/10.1007/978-3-031-92353-1_5

Hans J. Specht's charisma and "straightforward thinking" were not expressed through grand gestures or commanding speeches but through his ability to cut straight to the essence of the matter, inspire confidence, evoke curiosity, and lead by example. His decades of work at institutions such as the University of Heidelberg, CERN, GSI, and both TUM and LMU in Munich left an indelible mark on the fields of atomic, nuclear, and ultra-relativistic heavy-ion physics. More importantly, over five decades, he guided his colleagues and students—his "disciples"—not just toward scientific discoveries but toward becoming better scientists, thinkers, and individuals. His unique ability to unite people around shared goals, no matter how ambitious, remains one of his greatest legacies.

While this volume can only capture a small part of Hans's immense contributions to physics and society, we hope it serves as a source of inspiration for future generations of scientists. Hans J. Specht believed deeply in the power of curiosity, collaboration, and a commitment to the greater good. His life's work echoes Albert Einstein's sentiment: "*I get most joy in life out of music.*" For Hans, the harmony and creativity he found in music mirrored the very essence of his scientific pursuits. It is in that same spirit of passion and dedication that we share his story.

As Hans J. Specht might have said, life is not only about discovery but also about what we give to others along the way. With that spirit in mind, we conclude this memoir, trusting that his legacy will continue to inspire and guide future generations, just as he inspired us all.

With gratitude.

Co-editors: Sanja Damjanovic, Volker Metag, and Jurgen Schukraft

Open Access This chapter is licensed under the terms of the Creative Commons Attribution 4.0 International License (http://creativecommons.org/licenses/by/4.0/), which permits use, sharing, adaptation, distribution and reproduction in any medium or format, as long as you give appropriate credit to the original author(s) and the source, provide a link to the Creative Commons license and indicate if changes were made.

The images or other third party material in this chapter are included in the chapter's Creative Commons license, unless indicated otherwise in a credit line to the material. If material is not included in the chapter's Creative Commons license and your intended use is not permitted by statutory regulation or exceeds the permitted use, you will need to obtain permission directly from the copyright holder.

Acknowledgments

We thank the family of Hans J. Specht—Adelheid, Martin, Katja, and Michael Specht—for supporting this work and for their generous sponsorship of this book.

A special thanks to Georg Wolschin of the University of Heidelberg for his expert guidance and unwavering support throughout this Book's journey—from its initial concept to the final pages.

We would also like to express our thanks to Hisako Niko, the editor at Springer, for her helpful support and guidance throughout the editorial process. Her expertise and commitment have played a key role in bringing this book to publication.

The Authors

Michael Albrow

Michael Albrow obtained a Ph.D. from the University of Manchester in 1969. After a CERN Fellowship, he was a scientist at the Rutherford Laboratory before becoming a professor at Stockholm University in 1989. He was the spokesperson of the Axial Field Spectrometer at the ISR and Chair of the CERN PSSC Committee, and is now a Fellow of the Institute of Physics and the American Physical Society. Albrow went to Fermilab in 1991 and participated in the discoveries of the top quark and the Higgs boson. As a scientist emeritus, he does science writing for the general public, for science and art exhibitions, and for KV265 "Science and Symphony" concerts.

Peter Armbruster

Peter Armbruster was a pioneering German physicist, co-discovering the superheavy elements 107 (bohrium), 108 (hassium), 109 (meitnerium), 110 (darmstadtium), 111 (roentgenium), and 112 (copernicium). He earned his Ph.D. in 1961 under Heinz Maier-Leibnitz at TU Munich. His research focused on nuclear fission, heavy-ion interactions, and atomic physics. He held positions at the Research Centre Jülich (1965–1970) and GSI as senior scientist (1971–1996), and was Research Director at the Institute Laue-Langevin in Grenoble (1989–1992). He was also a professor at the University of Cologne and TU Darmstadt. Among his honors are the Max Born Medal and the Stern–Gerlach Medal.

Ulrich Charisius

Ulrich Charisius is a master piano and harpsichord maker based in Heidelberg, Germany. He is well known for his expertise in piano tuning for professional performances, including those of the renowned pianist Grigory Sokolov.

Sanja Damjanovic

Sanja Damjanovic received her Ph.D. in high-energy nuclear physics from the University of Heidelberg in 2002 under the supervision of Professor Hans J. Specht. Since 1999, she has been working at CERN and GSI-FAIR. Beyond her scientific career, she served as the Minister of Science of Montenegro (2016–2020). In 2017, she initiated SEEIIST, a pan-European research infrastructure for cancer therapy and research, and led its Intergovernmental Steering Committee until 2021. She has a permanent position at GSI.

Jürgen Debus

Jürgen Debus studied physics and medicine at the University of Heidelberg, where he received his Dr. rer. nat in 1991 and Dr. med in 1992. He is a distinguished German radiologist and radiation therapist, serving as Medical Director of the Heidelberg Ion Beam Therapy Center (HIT) since its inception and as a professor at the University of Heidelberg. He is also the Chairman of the Executive Board and Medical Director of the University Hospital Heidelberg. Professor Debus has made pioneering contributions to ion beam cancer therapy, notably as Medical Coordinator of the heavy-ion therapy project at GSI since 1994, where the first patients were treated with C-ions. His career began at the DKFZ, followed by a fellowship in proton therapy at the Massachusetts General Hospital, Boston. He later earned his habilitation and quickly became the Medical Director of the Department of Clinical Radiology/Radiation Therapy at the University Hospital Heidelberg. Professor Debus has held key roles in scientific societies and committees, including Director of the National Center for Tumor Diseases (NCT) and Chairman of the Scientific Council of the DKFZ.

Hans Günter Dosch

Hans Günter Dosch was born in Heidelberg in 1936 and studied physics in Heidelberg and Paris. He earned his Ph.D. in theoretical physics (supervisors J.H.D. Jensen, B. Stech) in 1963. After holding positions in Heidelberg, CERN, MIT, and KIT, he was appointed professor of theoretical physics in Heidelberg 1969, where he has been emeritus since 2002. His main research subject is phenomenology of strong interactions, and now holographic models for strong interactions. He has collaborated extensively with institutions in Montpellier, ITEP (Moscow), Pisa, Rio de Janeiro, Lanzhou, Stanford (Passadena), and San Jose (Costa Rica).

Axel Drees

Axel Drees is a distinguished professor at Stony Brook University. He earned his Ph.D. from Heidelberg University in 1989 for pioneering work on creating the QGP under Hans Specht's mentorship. Remaining at Heidelberg, he contributed to the CERES experiment at CERN SPS, which provided the first experimental evidence of thermal radiation from the QGP. After completing his habilitation in 1997, Drees joined the nuclear physics faculty at Stony Brook, where he played a key role in the discovery of jet quenching with the PHENIX experiment. He continues work on heavy-ion research with PHENIX and sPHENIX, and has served Stony Brook as department chair, dean, and vice provost.

Hartmut Eickhoff

Hartmut Eickhoff received his Ph.D. in 1977 from the University of Münster and worked at GSI from 1980 to 2013, where, among other tasks, he headed the Accelerator Division for several years and also served as Managing Technical Director. During the GSI therapy project, he was responsible for the necessary technical adaptations to the GSI accelerator facility and, subsequently, for the design and implementation of the accelerator system for the Heidelberg Therapy Facility (HIT). Dr. Eickhoff received the Otto Hahn Award in 2000 and retired in 2013 after a distinguished career in accelerator development.

Charles Gale

Charles Gale is a Distinguished James McGill Professor in the Department of Physics at McGill University. Active in research for over four decades, his work in the field of high-energy nuclear science explores strongly interacting matter under extreme conditions of temperature and density. He has developed theoretical models describing the dynamics of relativistic heavy-ion collisions and contributing to the formulation of field theories at finite temperature. He has also conducted extensive research on electromagnetic radiation from high-energy nuclear interactions.

Klaus Dieter Gross

Klaus Dieter Gross studied physics at the University of Mainz, earning a diploma in atomic collisions with polarized electrons. He obtained his Ph.D. in nuclear solid-state physics from the Free University and the Hahn–Meitner Institute in Berlin. After a two-year postdoc, he moved to GSI Darmstadt as assistant to the Scientific Managing Director, first to Paul Kienle and then to Hans Specht. He was later promoted to Head of the Management Staff of GSI, a position he held until 2022.

Dietrich von Harrach

Dietrich von Harrach received his Ph.D. from the Technical University of Munich in 1974. From 1974 to 1983, he worked as a research assistant at the University of Heidelberg, where he also completed his habilitation in 1983. From 1983 to 1990, he worked at CERN as a research associate at the Max Planck Institute for Nuclear Physics in Heidelberg. As a scientific associate from 1986 to 1988, he contributed to key projects at CERN like the discovery of the EMC effect. Since 1990, he has been a professor at the University of Mainz, focussing his research on parity-violating electron scattering. From 1991 to 1994, he was the executive director of the Institute of Nuclear Physics. After his retirement in 2010, he began to work on discrete quantum coherent spaces and entropic gravitation.

Volker Koch

Volker Koch received his Ph.D. in 1990 from the University of Giessen in Germany. After four years as a postdoc and visiting assistant professor at SUNY Stony Brook he moved to the Lawrence Berkeley National Laboratory (LBNL), first as a Divisional Fellow and then as Senior Scientist. He has served as Deputy Director for Low Energy Nuclear Science, Program Head for Nuclear Theory, and Interim Director of the Nuclear Science Division. He is a supervisory editor of *Nuclear Physics A* and a Fellow of the American Physical Society. His research focuses on the properties of strongly interacting matter.

Ewald Konecny

Ewald Konecny was a pioneering figure in both academic research and medical technology. He earned his Ph.D. in 1963 from the Technical University of Munich with a thesis on the mass spectrometric separation of nuclear fission fragments at

Germany's first research reactor. In 1967, he habilitated at the University of Giessen. His career in industry began in 1975 at Drägerwerk AG, where he developed innovative corporate strategies and led research in medical technology. By 1981, he became head of development, driving progress in sensors, microprocessors, and integrated workstations for anesthesia and ventilation equipment.

Thomas Kühl

Thomas Kühl studied physics at Johannes Gutenberg University in Mainz and received his Ph.D. based on a thesis conducted at CERN. After returning from a post-doctoral position at MIT, he became the leader of the laser spectroscopy group at GSI. His scientific focus included the spectroscopy of radioactive isotopes at the accelerator and, later, highly charged ions in the ESR storage ring. Influenced by Rudolf Bock and Hans Specht, he was attracted to the thematic of fusion energy and spent a year at the Nova Petawatt installation at Livermore in preparation for a high-energy laser project at GSI.

Volker Metag

Volker Metag studied physics at TU Berlin and the University of Heidelberg, where he received his Ph.D. in 1970. He held postdoc positions at the Max Planck Institute for Nuclear Physics (Heidelberg), the Niels Bohr Institute (Copenhagen), and the University of Washington (Seattle). Together with Dietrich Habs he received the Prize of the German Physical Society in 1978 for the spectroscopy of fission isomers. From 1982 until his retirement in 2012, he was professor of experimental physics at the University of Giessen. His research field is hadron and nuclear physics. He is a member of several international research collaborations and has served on numerous scientific advisory boards and as co-editor of *Physics Letters B*. From 1993 to 1998, he was research director of GSI, Darmstadt, and was involved in the preparation of the FAIR project.

Shoji Nagamiya

Shoji Nagamiya is currently a Visiting Research Director at RIKEN and a Diamond Fellow at KEK. He received his Ph.D. from Osaka University. Nagamiya began his career at the Lawrence Berkeley National Laboratory (LBL) (1973–1982), working primarily on the BEVALAC experiment. He then became Associate Professor at the University of Tokyo and a professor at Columbia University (1982–1997), contributing to AGS and RHIC, including as spokesperson for AGS E802 and

PHENIX. In 1997, he joined KEK as a professor, later managing JAEA's planning and development of J-PARC. He served as the Director of J-PARC until 2012. Between 2013 and 2018, he was a Research Advisor at RIKEN.

Thomas Nilsson

Thomas Nilsson is a Professor at Chalmers University of Technology in Gothenburg and the Scientific Managing Director of GSI and FAIR since December 2024. He is a member of the Royal Swedish Academy of Sciences, which decides on the Nobel Prize in Physics and Chemistry. He did his Ph.D. at TU Darmstadt and Chalmers. He was a physics coordinator at CERN's ISOLDE facility (1998–2004) and held research positions at TU Darmstadt and Chalmers, becoming a professor in 2009 and leading the physics department there from 2017. His research focuses on fundamental interactions manifested in the structure of exotic nuclei, with experiments at CERN and GSI/FAIR. He has served on the FAIR Council, GSI Supervisory Board, and advisory bodies at TRIUMF and RIKEN.

Ralf Rapp

Ralf Rapp is a Professor of Physics at Texas A&M University. He earned his Ph.D. in theoretical nuclear and hadron physics from the University of Bonn in 1996. His research focuses on quantum chromodynamics, heavy-ion collisions, and the quark–gluon plasma. He has held positions at SUNY Stony Brook, NORDITA, and Texas A&M. He has received the U.S. National Science Foundation CAREER Award (2004), the Bessel Award from the Humboldt Foundation (2007), and was named a Fellow of the American Physical Society in 2014. Rapp has an extensive publication record and has supervised numerous graduate students and postdoctoral researchers.

Carlo Rubbia

Carlo Rubbia, Nobel Laureate in Physics (1984), is renowned for discovering the W and Z particles at CERN in 1983. After earning his Ph.D. at the University of Milan, he conducted research at Columbia University before returning to Rome and joining CERN in 1960. Appointed Higgins Professor at Harvard University in 1970, he balanced teaching with research at CERN for 18 years. As CERN's Director-General (1989–1993), he was instrumental in developing the Large Hadron Collider and supporting the early adoption and global use of the World Wide Web. His contributions extend beyond particle physics, leading groundbreaking projects in accelerator technology, nuclear energy, and neutrino detection, such as the ICARUS experiment and the energy amplifier concept. A visionary in sustainable energy, he served as President of ENEA (1999–2005) and has advised global energy policy as a Senator for Life since 2013. Rubbia has received numerous prestigious honors, including Cavaliere di Gran Croce OMRI (1984) and, that same year, the Golden Plate Award. He holds 38 honorary degrees and is a member of various scientific academies, including the Pontifical Academy of Sciences. He has been further honored by the naming of asteroid 8398 Rubbia.

Helmut Satz

Helmut Satz is a German theoretical physicist whose main area of research is the study of strongly interacting matter. Until his retirement, he was a professor of physics at the University of Bielefeld. He studied at Michigan State University and the University of Hamburg, where he earned his Ph.D. in 1963. During periods of extensive leave, he was also a staff member at Brookhaven National Laboratory and CERN, as well as Gulbenkian Professor at the Technical University of Lisbon. Satz focused particularly on the physics of the quark–gluon plasma (QGP). One such signature, developed by Satz and Tetsuo Matsui in 1986, is the suppression of the J/ψ particle.

Horst Schmidt-Böcking

Horst Schmidt-Böcking studied physics in Würzburg and Heidelberg, receiving his Ph.D. in nuclear physics under Otto Haxel in 1969. After positions in Frankfurt and at the Hahn–Meitner Institute, he became professor at the University of Frankfurt, where he remained until retiring in 2004. His work on the spectroscopy of slow recoil ions led to the COLTRIMS reaction microscope, enabling the visualization of ultrafast reaction dynamics in atomic and molecular reactions with sub-attosecond resolution. He has received the Davisson–Germer Prize (2008) and the Stern–Gerlach Medal (2010).

Peter Schneider

Peter Schneider, a physicist and church musician, is an Associate Professor at Graz University. He studied both subjects in Freiburg and Heidelberg, earning diplomas in 1994 and 1996, and completed his Ph.D. in Physics (2000) under H.J. Specht and H.G. Dosch. In 2003, he founded the "Music and Brain" research group at Heidelberg University Hospital, collaborating with institutions in Mannheim, Basel, and Riga. Following his 2012 habilitation on the neural foundations of sound perception, he led several BMBF- and DFG-funded projects. Since 2021, he has relocated his research group to Graz, and founded the "Heidelberger Hörakademie."

Reinhold Schuch

Reinhold Schuch received his Ph.D. in solid state physics from the University of Göttingen in 1975, then moved to the University of Heidelberg to work on heavy-ion atomic collision physics. He made extended research visits to RIKEN and NIF in Japan, as well as at the Brookhaven Tandem Laboratory. From 1986 to 1987, he took up employment as a researcher at Oak Ridge National laboratory. He later became a professor at the Manne Siegbahn Institute in Stockholm and, in 1991, a professor at Stockholm University. He is also a member of the Royal Swedish Academy of Sciences.

Jurgen Schukraft

Jurgen Schukraft received his Ph.D. in nuclear physics from the University of Heidelberg in 1983 under the supervision of H.J. Specht. Joining CERN in 1984, he worked on proton–proton collisions at the ISR, and later, heavy-ion experiments at the SPS and Brookhaven's AGS. A founding member of the ALICE experiment at the LHC, he served as its first spokesperson from 1991 to 2010. After his retirement from CERN in 2018, he held a Distinguished Professorship at CCNU Wuhan and is currently Adjunct Professor at Yale University and Affiliated Professor at the Niels Bohr Institute.

Hans Joachim Specht

Born in 1936 in Unna, Westphalia, Germany, Hans Joachim Specht studied physics at LMU München, TU München, and ETH Zürich, earning his Ph.D. in 1964 under the mentorship of Heinz Maier-Leibnitz at the FRM reactor and TU München. Following research in Canada at the AECL Nuclear Laboratories, he returned to LMU as professor. In 1973, he was appointed professor at the University of Heidelberg, where he initially conducted research at MPI Heidelberg and GSI Darmstadt, later establishing a leading high-energy heavy-ion research group with strong links to CERN. A pioneer of CERN's ultra-relativistic heavy-ion program, Specht contributed to four major experiments: R807/808 at the ISR, HELIOS/NA34 (spokesperson), NA45/CERES (founder and spokesperson), and NA60. His early conceptual design of a heavy-ion experiment at the LHC, along with his advocacy for institutional support, were instrumental in shaping ALICE into one of CERN's flagship experiments. From

1992 to 1999, he served as Scientific Managing Director of GSI, where he initiated Europe's first successful cancer therapy using carbon ions. A member of the Heidelberg Academy of Sciences since 2000, he became Professor Emeritus in 2004. His research spanned atomic physics (quasi-molecules), nuclear fission (shape isomers), high-energy physics (quark–gluon plasma), and neuroscience (music perception and early brain processing). Over the course of his career, he supervised more than 100 Ph.D. and diploma students. Hans J. Specht passed away on 20 May 2024.

Dinesh Kumar Srivastava

Dinesh Kumar Srivastava graduated from Allahabad University in 1970 and joined the Training School at the Bhabha Atomic Research Center, Mumbai, before moving to the Variable Energy Cyclotron Project in 1971. He retired in 2016 as a distinguished scientist and director of the Variable Energy Cyclotron Center, Kolkata. His key contributions include studies of the density dependence of nuclear forces, Coulomb and nuclear breakup of light nuclei, photon and dilepton radiation from QGP, heavy quark energy loss in QGP, and the parton cascade model for early-stage heavy-ion collisions. Today, he is mainly engaged in science outreach.

Andreas Tauschwitz

Andreas Tauschwitz studied physics at the University of Mainz and Darmstadt, graduating with an engineering degree in physics in 1990. For his Ph.D., he joined the Plasma Physics Department at GSI. After completing his doctorate, he worked as a postdoctoral researcher at prestigious institutions including the University of California, Berkeley, Forschungszentrum Karlsruhe, and the University of Erlangen. Finally, he returned to GSI, where he worked on the installation of the PHELIX laser and the preparation of plasma physics facilities for the FAIR project.

Gianluca Usai

Gianluca Usai is a full professor at the University of Cagliari, where he earned his Ph.D. in 1995, specializing in experimental high-energy physics. His academic career progressed from an INFN postdoctoral researcher to associate professor, and later, full professor. A leading expert in quark–gluon plasma, he has made significant contributions to major experiments, including NA50, NA60, and ALICE at CERN. Since 2005, he has served as the NA60 spokesperson, pioneering the development of radiation-hard silicon pixel detectors. Currently, he leads the proposal for CERN's future NA60+ experiment, focusing on electromagnetic probes of QCD matter at SPS energies. His work bridges fundamental QCD research with cutting-edge detector innovation.

Jochen Wambach

Jochen Wambach is a German theoretical nuclear physicist who has been a professor at the Technical University of Darmstadt since 1986. He focuses primarily on theoretical nuclear and hadron physics. Wambach received his Ph.D. from the University of Bonn. In 1996, he succeeded Friedrich Beck as a professor at TU Darmstadt. He became a Fellow of the American Physical Society in 2003. In 2015, he was inducted into the Academia Europaea. Since 2007, he has been an editor of the *European Journal of Physics A* and *Physical Review Letters*. Wambach served as Director of ECT* Trento from 2016 to 2020.

Wolfram Weise

Wolfram Weise received his Ph.D. from Friedrich Alexander University of Erlangen-Nürnberg in 1970 and completed his habilitation in 1974. He worked as a Research Associate at the State University of New York at Stony Brook (1973–1975) and as a Scientific Associate at CERN, Geneva (1975–1976). Weise then became a professor of physics at the University of Regensburg (1976–1994) and later a professor of theoretical physics at TU Munich (1994–2012). He served twice as Director of ECT* in Trento (2001–2004 and 2012–2015) and has been a professor emeritus at TUM since 2012, holding the title of TUM Emeritus of Excellence since 2015.

GPSR Compliance

The European Union's (EU) General Product Safety Regulation (GPSR) is a set of rules that requires consumer products to be safe and our obligations to ensure this.

If you have any concerns about our products, you can contact us on

ProductSafety@springernature.com

In case Publisher is established outside the EU, the EU authorized representative is:

Springer Nature Customer Service Center GmbH
Europaplatz 3
69115 Heidelberg, Germany

www.ingramcontent.com/pod-product-compliance
Lightning Source LLC
LaVergne TN
LVHW011000250326
834688LV00003B/31